# Lecture Notes in Computer Scie

*Commenced Publication in 1973*
Founding and Former Series Editors:
Gerhard Goos, Juris Hartmanis, and Jan van Leeuwe

T0238187

Hans Dobbertin   Vincent Rijmen
Aleksandra Sowa (Eds.)

# Advanced
# Encryption Standard –
# AES

4th International Conference, AES 2004
Bonn, Germany, May 10-12, 2004
Revised Selected and Invited Papers

 Springer

Volume Editors

Hans Dobbertin
Ruhr-University of Bochum
Cryptology and IT Security Research Group
Universitätsstrasse 150, 44780 Bochum, Germany
E-mail: Hans.Dobbertin@ruhr-uni-bochum.de

Vincent Rijmen
Graz University of Technology
Institute for Applied Information Processing and Communications (IAIK)
Inffeldgasse 16a, 8010 Graz, Austria
E-mail: vincent.rijmen@iaik.tugraz.at

Aleksandra Sowa
Ruhr-University of Bochum
Horst Görtz Institut für Sicherheit in der Informationstechnik
Universitätsstrasse 150, 44780 Bochum, Germany
E-mail: Aleksandra.Sowa@hgi.ruhr-uni-bochum.de

Library of Congress Control Number: 2005928447

CR Subject Classification (1998): E.3, F.2.1-2, I.1.4, G.2.1

ISSN     0302-9743
ISBN-10  3-540-26557-0 Springer Berlin Heidelberg New York
ISBN-13  978-3-540-26557-3 Springer Berlin Heidelberg New York

Springer is a part of Springer Science+Business Media

springeronline.com

© Springer-Verlag Berlin Heidelberg 2005
Printed in Germany

Typesetting: Camera-ready by author, data conversion by Scientific Publishing Services, Chennai, India
Printed on acid-free paper     SPIN: 11506447     06/3142     5 4 3 2 1 0

# Preface

This volume comprises the proceedings of the 4th Conference on Advanced Encryption Standard, 'AES — State of the Crypto Analysis,' which was held in Bonn, Germany, during 10–12 May 2004.

The conference followed a series of events organized by the US National Institute of Standards and Technology (NIST) in order to hold an international competition to decide on an algorithm to serve as the Advanced Encryption Standard (AES). In 1998, at the first AES conference (AES 1), 15 different algorithms were presented, discussed, reviewed and verified. A second conference was organized in April 1999, and by August 1999 only five candidates were still in the running: MARS, RC6, Rijndael, Serpent and Twofish. After a further conference devoted to verification, testing and examination of the candidate algorithms in order to prove their performance and security, one winning algorithm remained. The encryption scheme Rijndael, designed by the Belgian cryptographers Joan Daemen and Vincent Rijmen, was selected in 2000 to become the successor to the famous DES (Data Encryption Standard) and it is now the Advanced Encryption Standard.

Like DES before it, AES is going to become a de facto world standard for the encryption of data. The security of Internet applications, for instance, is already depending today and, in view of the increasing implementation, will depend in future even more on AES. Analysis of the cryptographic strength of AES belongs therefore certainly to the most important topics in cryptology. A recent key recovery approach, by solving a complicated system of quadratic equations, which is due to Courtois and others, has caused a big debate. Previously, approaches of this kind were considered as purely theoretical, and hopeless in practice. The big unanswered question is whether the addition of newly proposed techniques has changed or can change this situation.

Four years after the National Institute of Standards and Technology chose Rijndael to be the Advanced Encryption Standard, leading experts and scientists from all over the world were invited to discuss — critically but constructively — the strengths and weaknesses of Rijndael, and to look for solutions that will make it a strong information encryption formula for the next two, five, ten, or maybe dozens of years. The intentions of the AES4 conference organizers were to present the most recent ideas and results on the cryptanalysis of the AES, and to stimulate future research on the important open questions about the perspectives and limits of new cryptanalytic approaches.

The response to the conference was excellent. Ten submission were selected for presentation. The programme included six keynote addresses (invited talks), given by Yvo Desmedt from Florida State University, Vincent Rijmen from the IAIK, Graz University of Technology and Cryptomathic, Carlos Cid from Royal Holloway, University of London, Nicolas T. Courtois from Axalto Smart Cards,

Jean-Charles Faugère from the University of Paris VI/INRIA, France, and John Kelsey from the National Institute for Standards and Technology. As a novum, AES4 introduced for the first time a closing panel discussion on the future of Rijndael and cryptography, moderated by Peter Welchering from the German Scientific Press Conference. Researchers took the opportunity to present their opinions and suggestions on the cipher weaknesses, known and unknown attacks, and the future of their work. John Kelsey remarked that most of the practical problems are usually other than the weaknesses of a cipher. Nevertheless, as Nicolas T. Courtois argued, there is still 'plenty of work' to do. Carlos Cid and Vincent Rijmen emphasized the necessity to make the current research transparent, to make it popular and understandable and to let other people know 'what we are talking about' (Vincent Rijmen).

We would like to thank Aleksandra Sowa, the Managing Director of the Horst Görtz Institute (HGI) for IT security at the Ruhr University of Bochum. She did an excellent job as General Chair by organizing the AES4 conference with the help of our young colleagues from the Chair for IT Security and Cryptology (CITS).

We are also grateful to NIST and Cryptomathic for supporting this event, and, last but not least, we would like to thank all the committee members for their work.

April 2005                                      Hans Dobbertin and Vincent Rijmen

# Organization

AES4 was organized by the Ruhr University of Bochum, in cooperation with the Graz University of Technology and NIST.

## General Chair

Aleksandra Sowa      Horst Görtz Institute, Ruhr University Bochum

## Program Co-chairs

Hans Dobbertin      Horst Görtz Institute, Ruhr University Bochum
Vincent Rijmen      Graz University of Technology

## Program Committee

Don Coppersmith      IBM
Nicolas T. Courtois      Axalto Smart Cards
Lars R. Knudsen      Technical University of Denmark
Matt Robshaw      Royal Holloway, University of London

## Sponsoring Institutions

Cryptomathic A/S, Århus
NIST

# Table of Contents

# Other Topics

# The Cryptanalysis of the AES - A Brief Survey

Hans Dobbertin[1], Lars Knudsen[2], and Matt Robshaw[3]

[1] Cryptology and IT Security Research Group,
Ruhr-University of Bochum, Germany
Hans.Dobbertin@ruhr-uni-bochum.de
[2] Department of Mathematics,
Technical University of Denmark,
DK-2800 Lyngby, Denmark
Lars.R.Knudsen@mat.dtu.dk
[3] France Télécom Research and Development,
38–40 rue de Général-Leclerc, 92794 Issy Moulineaux, France
Matt.Robshaw@francetelecom.com

**Abstract.** The Advanced Encryption Standard is more than five years old. Since standardisation there have been few cryptanalytic advances despite the efforts of many researchers. The most promising new approach to AES cryptanalysis remains speculative, while the most effective attack against reduced-round versions is older than the AES itself. Here we summarise this state of affairs.

## 1   Introduction

In January 1997 the National Institute of Standards and Technology (NIST) initiated the search for a replacement for the *Data Encryption Standard* (DES) [28]. The requirements for the new standard, to be called the *Advanced Encryption Standard* (AES), were that it should be:

- a 128-bit block cipher with the choice of three key sizes of 128, 192, respectively 256 bits,
- a public and flexible design,
- at least as secure as two-key triple-DES, and
- available royalty-free worldwide.

At the conclusion of this standardisation effort, with many man-years of cryptanalytic and implementation expertise provided from around the world, *Rijndael*, developed by Joan Daemen and Vincent Rijmen [11], was a popular choice to become the AES. In November 2001 the AES effort came to its conclusion with the publication of FIPS 197 [29], and today the AES is fast becoming a vital component of the digital infrastructure.

The proceedings of the Fourth AES Conference that follow in this volume reflect ongoing research efforts into the security and performance of the AES. In this short article, we briefly review some promising – but unsuccessful – attempts to compromise this elegant cipher.

H. Dobbertin, V. Rijmen, and A. Sowa (Eds.): AES 2004, LNCS 3373, pp. 1–10, 2005.

## 2    AES Design

The AES has been described so often and is, by now, so familiar that a brief overview of the AES design will suffice for our purposes.

The AES is a classic *substitution/permutation* or SP-network that requires 10, 12, or 14 rounds of encryption; the exact number depending on the length of the key. The AES is byte-oriented and heavily reliant on operations in the field $GF(2^8)$. Conceptually, the AES is best described with the sixteen bytes of the 128-bit input block $a_0 a_1 \ldots a_{14} a_{15}$ being arranged in a $(4 \times 4)$ matrix of bytes:

| $a_0$ | $a_4$ | $a_8$ | $a_{12}$ |
|---|---|---|---|
| $a_1$ | $a_5$ | $a_9$ | $a_{13}$ |
| $a_2$ | $a_6$ | $a_{10}$ | $a_{14}$ |
| $a_3$ | $a_7$ | $a_{11}$ | $a_{15}$ |

Using the nomenclature of FIPS 197, a typical round of the cipher uses the following operations, "SubBytes", "ShiftRows", "MixColumns" and "AddRoundKey". The final round has a slightly different form and omits the MixColumns operation.

Encryption begins with an AddRoundKey operation, then computation continues for a given number of rounds, with each round using the four operations taken in the order above. In SubBytes each byte is replaced by a byte from an invertible S-box. In ShiftRows the rows (of bytes) are shifted a number of byte positions to the left. The top row is not shifted, the second row is shifted by one position, the third by two, and the fourth row by three. In MixColumns the four bytes in each column are mixed by pre-multiplying the four-byte vector by a fixed, invertible, $(4 \times 4)$-matrix over $GF(2^8)$, that is derived from an MDS code. MixColumns has the property that if two input vectors differ in $s$ bytes, then the output vectors differ in at least $5 - s$ bytes, where $1 \leq s \leq 4$. Each round closes with AddRoundKey where 16 round-key bytes are exclusive-or'ed to the 16 data bytes. Each round uses all four operations except the last round when the operation MixColumns is omitted. We refer to [29] for more details on this and other aspects of the algorithm.

The key schedule for the AES is relatively simple. It takes the user-supplied key of 16, 24, respectively 32 bytes and returns what is called an ExpandedKey of $16 \times 11$, $16 \times 13$, and $16 \times 15$ bytes respectively. The details can be found in [13, 29].

## 3    The Components

By design, Rijndael, and therefore by extension the AES, is a very structured cipher. This very clean structure has at least two attractive consequences:

1. It is possible to provide a simple explanation for the intended effect of each cipher component. The most striking consequence is that we can derive solid reassurance for the resistance of the AES to basic differential [4] and linear [23] cryptanalytic attacks.
2. The implementor is provided with a wide range of implementation options. This is evidenced by the attractive performance profile of the AES across a wide range of environments.

We will explore the first consequence in this article.

### 3.1    The S-Box

The cryptographic strength or weakness of the AES depends strongly on the choice of S-box. While it is likely that we would view the S-box as a single entity, it has three distinct components; inversion over $GF(2^8)$ which is naturally augmented to handle the zero input, transformation by a $GF(2)$-linear map $L$, and addition of a constant $c = $ 0x63. Thus, up to a $GF(2)$-affine modification, the S-box $S(x)$ of the AES is the inversion in the multiplicative group of $GF(2^8)$:

$$S(x) = A(1/x) \quad \text{(with the convention } 1/0 = 0), \tag{1}$$

where $A(x) = L(x) + c$ is a $GF(2)$-affine permutation of $GF(2^8)$.

The cryptographic advantages of $1/x$ on $GF(2^n)$ have been known for some time. It realizes the best known properties of bijective S-box constructions with respect to the following properties:

**Degree.** *All S-box component functions (i.e. non-zero linear combinations of Boolean coordinate functions of the S-box) have degree $n - 1$.*
   The degree of all non-zero component functions of a non-constant *power function* $x^d$ is the Hamming weight of the binary representation of the remainder of $d$ modulo $2^n - 1$. Thus the maximal degree $n - 1$ is achieved if, and only if, up to cyclotomic equivalence, $d = -1 = 2^n - 2 = 2(1 + 2 + 2^2 + ... + 2^{n-2}) \bmod (2^n - 1)$. On the other hand it is well known that each component function of a *one-to-one* S-box has at most degree $n - 1$.
**Resistance to linear attack.** *Low correlation between S-box component functions and affine Boolean functions.*
   The absolute value of the correlation between any non-zero component function of $1/x$ and any affine Boolean function is bounded by $2^{-n/2-1}$ for even $n$. This can be shown by using the famous Hasse bound for the number of points on elliptic curves. It is an open problem whether this bound can be improved. We mention that for odd $n$, the bound $2^{-n-1/2}$ is attained by $1/x$ and this is known to be optimal.
**Resistance to differential attack.** *The designer's dream "for each prescribed input difference one can derive no information about the S-box output difference" is almost achieved.*
   For characteristic 2, differences coincide with sums. Thus the number of possible output differences for pre-scribed input difference is at most $2^{n-1}$. If this

bound is achieved then the S-box is called *almost perfect nonlinear (APN)*, and in this case each output difference is attained precisely two times. If $n$ is odd then $1/x$ is APN, while $2^{n-1} - 1$ is the number of output differences for even $n$. The latter is due to the fact that $1/x$ is linear on GF(4). It is not known if there is any APN *one-to-one* S-box for even $n$.

These properties of inversion are preserved under affine modifications and are therefore valid for the S-box of the AES. The net result is an exceptional resistance to differential and linear cryptanalysis. In [11] it is shown that any four-round differential characteristic has a probability of less than $2^{-150}$ and that any four-round linear characteristic holds with a correlation less than $2^{-75}$. These bounds are sufficient to conclude that the basic attacks based on differential and linear cryptanalysis will not succeed against the AES.

While the resistance of the AES to advanced attacks or those using differentials and/or linear hulls remains open, there have been a series of results that explore these issues [8, 18, 19, 20, 21, 30, 31, 7, 32, 33, 34, 5]. However there seems little chance of a major breakthrough in this direction.

### 3.2     Rearranging Components

While the structure of Rijndael received cryptanalytic attention during the AES process, (see Section 4) it was only at the tail end of that process that a different kind of observation began to be explored. These observations are based on alternative representations of components, or the entireity, of the AES. Some researchers have considered a continued fraction representation of AES encryption [16] while others have considered the concept and implications of *dual Rijndaels* [3, 35]. Other observations have been concerned with the way AES operations are presented [25, 26].

Clearly, operations such as `SubBytes` and `ShiftRows` trivially commute with one another. Indeed, properties such as these were used by the AES designers to show how AES decryption could be written in a form that more closely resembled encryption. However a more fundamental re-writing is also possible. While it is typical to take the S-box as a single entity, we have already observed that it consists of three separate components; the augmented inversion mapping $1/x$, the linear map $L$, and addition of the constant `0x63`. Concern about the algebraic simplicity of the inversion operation over $GF(2^8)$ lead the designers to introduce a mixing function (the linear map $L$) over $GF(2)$, while concern that the input 0 would be mapped to 0 through the two combined operations lead to the final addition of the constant.

Yet, it is instructive to view this package as the sequence of independent operations it truly is [25, 26]. It is then trivial to see that the parallel addition of sixteen constants `0x63` can be moved (unchanged) through the `ShiftRows` operation. It can also be moved (unchanged) through the `MixColumns` operation. We might therefore remove the addition of the constant from the encryption process entirely and, instead, consider it a minor addition to the key schedule. We can also view the sixteen parallel applications of the linear map $L$ as part of the diffusion layer that follows. While making the diffusion layer slightly more complicated than that given in the standard description, this separation of the components of the AES yields a more unified functional description.

The value of such rewriting has been questioned [12], but it does provide some additional perspectives on the AES structure. But while there is some interaction between this line of work and the aims of algebraic cryptanalysis (see Section 5) these different perspectives on the AES have yet to yield any practical cryptanalytic advance. Instead the most successful attacks on the AES are of an entirely different nature.

## 4 Structural Attacks

The most effective attacks on reduced-round variants of the AES are variants of the *Square* attack which is due to Knudsen. Since this attack was used against a predecessor [10] of the AES it was accounted for by the AES designers [11].

In this attack we take a set of 256 plaintexts where the first byte takes all possible values. The other 15 bytes of the input can take any value but the same value in a given byte position must be used across all 256 texts. We will describe a set of texts that have this property as an *integral*. Imagine one begins an AES-round with such an integral. In the following we shall denote the byte-position containing a variable value with an "$a$" (for "all"). Consider the actions of SubBytes, ShiftRows, and MixColumns.

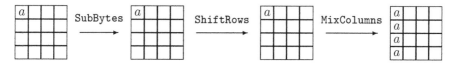

The AddRoundKey operation adds the same round key to each of the 256 texts in the integral, therefore any integral before AddRoundKey will yield an integral after. Consider a second round of transformation.

It follows that after two rounds of encryption and for each byte position, every possible value in a given byte position is taken once and only once in the set of 256 texts. Now consider a third round.

Here $s$ indicates that the sum of the texts in a particular byte can be determined (and in this case is equal to zero). The interesting part is what happened during the MixColumns operation. Before the operation, in each byte position the 256 values were a permutation of the values $0, \ldots, 255$. MixColumns combines four bytes to yield one byte in a linear way. This means that after the application of MixColumns every byte position will be balanced, that is, if we exclusive-or all 256 values in any single byte position we will get zero as a result. Note how this property, after three rounds of AES encryption does not depend on the details of the S-box nor on the value of the secret key.

Such three-round structures can be used to attack the AES reduced to six rounds (where the first round consists of AddRoundKey and the last round is without MixColumns). The structure is used over rounds two to four. Then by guessing four key bytes in the first round, four key bytes in the final application of AddRoundKey and one key byte in the second-last application of AddRoundKey, in total nine key bytes, one can compute a candidate value for the sum of the texts in one byte position after four rounds of encryption. For a structure of 256 plaintexts of the form above, this sum is known to be zero. In fact, there will be values of the nine key bytes that will return zero as the value of the sum by chance. So to eliminate false alarms, the attack needs to be repeated a few times to uniquely determine the correct key bytes. Once the nine key bytes have been found, we find the remaining twelve key bytes of the final application of AddRoundKey, after which the user-selected key can be derived. Taking advantage of some advanced observations, there is a more effective extension of this attack. This can be used to find the secret key with $6 \cdot 2^{32}$ chosen plaintexts in a time equivalent to $2^{44}$ encryptions and $2^{32}$ words of memory [15, 22]. There have also been some further extensions to the basic Square attack, but these require an explosive increase in the running time [15, 22].

Another kind of structural attack that has been described against the AES is sometimes referred to as a *collision* or *bottleneck* attack. These attacks require around $2^{32}$ plaintexts and exploit a three-round structure [17, 24]. These approaches can be used to attack AES reduced to seven rounds but the running time is almost the same as an exhaustive search for the key.

## 5   Algebraic Attacks

We saw in Section 3.1 that the S-box was carefully constructed around inversion in $GF(2^8)$. As a consequence, if we appeal to our earlier notation (1), then we have the implicit equation $A^{-1}(S(x))x = 1$ for $x \neq 0$. Thus there are eight *quadratic* equations that relate the bits of $S(x)$ and $x$. Of these eight equations seven holds always, while the eighth holds only when $x \neq 0$, that is, in 255 of 256 cases. In addition, another 32 quadratic equations can be derived since $xy = 1$ implies that

$$x^2 y = x, \; xy^2 = y, \; x^4 y = x^3, \text{ and that } xy^4 = y^3.$$

Each of these equations leads to eight quadratic equations on the bit level and all of these always hold. The resulting 39 quadratic equations turn out to be a base for the vector space, over GF(2), of *all* quadratic equations relating the input and output bits of the AES S-box.

In every round of the AES, a parallel block of several instances of the S-box is applied. But everything else that happens in the encryption/decryption procedure outside the S-boxes is GF($2^8$)-linear. These observations motivate a tempting idea. Suppose a plaintext block $P$ and the corresponding ciphertext block $C$ (or a collection of such pairs) were known. Could we recover the encryption key $K$ by establishing, and solving, a binary quadratic equation system?

Establishing such a quadratic equation system is easy. The bits of the expanded round keys, which we consider unknowns, are added bitwise (in byte blocks) to each S-box input. We introduce new variables for the input and output bits of each occurring S-box, and relate them by the quadratic equations mentioned above. The disadvantage is that this leads to a huge number of binary variables.

Everything in the AES encryption/decryption algorithm can be described equally over GF(2) or GF($2^8$) with two exceptions. The first exception is inversion over GF($2^8$); this makes it hard to work over GF(2). However, the inversion operation would be much easier if we could work over GF($2^8$). But at this point, the second exception comes into play. The GF(2)-linear mapping $L$, used to modify $1/x$ in the S-box, appears to prevent an algebraic attack with *quadratic* equations over GF($2^8$). Indeed, it was the intention of the AES designers to destroy the GF($2^8$)-structure by taking an affine modification of $1/x$. However, it was observed [27] that the GF(2)-linear mappings

$$L : \mathrm{GF}(2^n) \longrightarrow \mathrm{GF}(2^n)$$

are precisely those, which can be written as polynomials in the form

$$L(x) = \sum_{i<n} \alpha_i x^{2^n},$$

with uniquely determined $\alpha_i \in \mathrm{GF}(2^n)$ $(i < n)$. Thus the GF(2)-affine map $A$ can be represented as $A(x) = \sum_{i<8} a_i x^{2^i} + c$ with coefficients in GF($2^8$). This then allows AES encryption to be represented as a sparse system of quadratic equations over GF($2^8$) and there are strong reasons for expecting such a system to be easier to solve than the corresponding system over GF(2) [27].

There is much discussion and speculation about whether such algebraic attacks might ever be relevant. While some very positive views have been expressed [9], most researchers are more cautious. Without doubt, the need to find powerful elimination techniques for multi-variate equation systems is a significant topic in cryptography, though not only in the context of block ciphers. Some well known basic ingredients such as

 - linearization (substitution of monomials by single new variables) and
 - Buchberger's algorithm (computations of Gröbner bases),

have lead to algorithms such as XL (eXtended Linearization), XSL (eXtended Sparse Linearization), and also the algorithms F4 and F5 due to Faugère. Various web-links on this topic are available [9]. Unfortunately the complexity of these algorithms is closely related to very difficult problems in algebraic geometry and commutative algebra, and heuristics are very risky. A recent result leading to a re-evaluation of the XL algorithm exemplifies this [14].

The threat posed by algebraic attacks on the AES is difficult to quantify. As things stand, there is little belief that the XSL algorithm—which was explicitly formulated to work with the AES equation systems—will work as orginally hoped. That said, there is no intrinsic reason why new variants of XL or XLS might not work at some stage in the future.

Instead, the AES system of equations over either $GF(2)$ or $GF(2^8)$ is currently "best" solved [6] using Gröbner basis techniques such as Buchberger's algorithm or $F_4$. However there are many complications and there remains much to understand. Furthermore, recent experimental work [6] has shown that basic implementations of these algorithms are limited, and memory limitations thwart attempts to cryptanalyse even the most basic variants of the AES. However all is not lost for the cryptanalyst.

One frustrating aspect of the experimental work in [6] is that simple changes to the way the equation systems are presented can have an unpredictable effect on the solution time. Furthermore, equal-sized equation systems arising from two very different AES-variants can take very different amounts of time to solve. However, if equal-sized equation systems can take different times to solve, then the solving algorithm would seem to be taking advantage of hidden structure in one of the cases. Thus it might be hoped that additional research will lead to more efficient, potentially AES-specific, solution methods in the future.

There is currently no solid estimation for the effort of an algebraic attack against the AES. To break AES in practice remains completely elusive. However for those interested in a challenge, one example in the MYSTERY TWISTER 2005 [2] cryptographic challenge is dedicated to an attack on a small scale variant of the AES [6] with 64-bit keys.

## 6   Conclusions

The AES is a *de facto* world standard with an intended lifespan of 30 years and, as the successor to DES, has much to live up to. Yet, apart from the currently speculative threat of algebraic attacks, there are few results since standardisation that would lead anyone to seriously question the practical security of the AES. Hopefully, by the time of some future AES5 conference, the true extent of block cipher algebraic cryptanalysis will be much clearer, but at the time of writing the AES has reached its fifth birthday in very good shape.

# References

1. AES web site of ECRYPT:http://www.iaik.tu-graz.ac.at/research/krypto/AES/
2. MYSTERY TWISTER web site: http://www.mystery-twister.com
3. E. Barkan and E. Biham, *In how many ways can you write Rijndael?*, Proceedings of Asiacrypt 2002, Lecture Notes on Computer Science, vol. 2501, Springer-Verlag, Berlin – New York, 2002.
4. E. Biham and A. Shamir. Differential Cryptanalysis of the Data Encryption Standard. Springer Verlag, 1993.
5. A. Biryukov, *The boomerang attack on 5 and 6-round reduced AES,* in these proceedings, pp. 13–17.
6. C. Cid, S. Murphy and M. Robshaw, *Small Scale Variants of the* AES, Proceedings of Fast Software Encryption 2005, Lecture Notes on Computer Science, vol. Springer-Verlag, Berlin – New York, to appear; see http://www.isg.rhul.ac.uk/~ccid/publications.htm
7. J.H. Cheon, M. Kim, K. Kim, J.-Y. Lee, and S. Kang, *Improved impossible differential cryptanalysis of Rijndaeland Crypton,* In 3$^{rd}$ International Conference on Information Security and Cryptology (ICISC 2001), volume 2288 of *Lecture Notes in Computer Science* 2288, Springer-Verlag, Berlin – New York, 2001, pp. 39–49.
8. K. Chun, S. Kim, S. Lee, S. Sung, and S. Yoon, *Differential and linear cryptanalysis for 2-round SPNs,* Information Processing Letters 87, 2003, pp. 277–282.
9. N. Courtois: Is AES a secure cipher? http://www.cryptosystem.net/aes/
10. J. Daemen, L. Knudsen, and V. Rijmen, *The block cipher Square,* In Fast Software Encryption, 4th International Workshop, FSE'97, Haifa, Israel, January 1997, E. Biham(ed.), Lecture Notes in Computer Science, vol. 1267, Springer-Verlag, Berlin – New York, 1997, pp. 149–165.
11. J. Daemen and V. Rijmen, AES *Proposal: Rijndael. Version 2.0,* available via http://www.crsc.nist.gov.
12. J. Daemen and V. Rijmen, *Answers to "New Observations on Rijndael".* Archived via http://www.crsc.nist.gov.
13. J. Daemen and V. Rijmen, *The Design of Rijndael. AES - The Advanced Encryption Standard,* Springer-Verlag, Berlin – New York, 2002.
14. C. Diem, *The XL-algorithm and a conjecture from commutative Algebra,* Asiacrypt 2004, December 2004, Korea, Lecture Notes in Computer Science, to appear.
15. N. Ferguson, J. Kelsey, B. Schneier, M. Stay, D. Wagner, and D. Whiting, *Improved cryptanalysis of Rijndael,* Fast Software Encryption, 7th International Workshop, FSE 2000, B. Schneier (ed.), New York, April 2000, Lecture Notes in Computer Science, vol. 1978, Springer-Verlag, Berlin – New York, 2001, pp. 213–230.
16. N. Ferguson, R. Shroeppel, and D. Whiting, *A simple algebraic representation of the* AES*,* Selected Areas in Cryptography, SAC 2001, S. Vaudenay and A.M. Youssef (editors), Lecture Notes in Computer Science, vol. 2259, Springer-Verlag, Berlin – New York, 2001, pp. 103–111.
17. H. Gilbert and M. Minier., *A collision attack on 7 rounds of Rijndael,* 3$^{rd}$ Advanced Encryption Standard Candidate Conference, National Institute of Standards and Technology, April 2000, pp. 230–241.
18. S. Hong, S. Lee, J. Lim, J. Sung, and D. Cheon, *Provable security against differential and linear cryptanalysis for the spn structure,* Fast Software Encryption, 7th International Workshop, FSE 2000, B. Schneier (ed.), New York, April 2000, Lecture Notes in Computer Science, vol. 1978, Springer-Verlag, Berlin – New York, 2001, pp. 273–283.

19. L. Keliher, *Refined analysis of bounds related to linear and differential cryptanalysis for the AES* these proceedings, pp. 45–60.

20. L. Keliher, H. Meijer, and S. Tavares, *New method for upper bounding the maximum average linear hull probability for SPNs* Advances in Cryptology - EURO-CRYPT'01, Birgit Pfitzmann(ed.), Lecture Notes in Computer Science, vol. 2045, Springer-Verlag, Berlin – New York, 2001, pp. 420–436.

21. L. Keliher, H. Meijer, and S. Tavares, *Improving the upper bound on the maximum average linear hull probability for Rijndael,* In *Selected Areas in Cryptography, 8th Annual International Workshop, SAC 2001 Toronto, Ontario, Canada, August 16-17, 2001,* S. Vaudenay and A. M. Youssef (ed.), Lecture Notes in Computer Science, vol. 2259, Springer-Verlag, Berlin – New York, 2001, pp. 112–128.

22. S. Lucks, *Attacking seven rounds of Rijndael under 192-bit keys and 256-bit keys,* Proceedings of the $3^{rd}$ Advanced Encryption Standard Candidate Conference, National Institute of Standards and Technology, April 2000, pp. 215–229.

23. M. Matsui, *The First Experimental Cryptanalysis of the Data Encryption Standard* Advances in Cryptology - CRYPTO'94, Yvo Desmedt (ed.), Lecture Notes in Computer Science, vol. 839, Springer-Verlag, Berlin – New York, 1994, pp. 26–39.

24. M. Minier, *A three rounds property of the AES,* these proceedings, pp. 18–29.

25. S. Murphy and M. Robshaw, *New Observations on Rijndael.* August 7, 2000. Archived via http://www.crsc.nist.gov.

26. S. Murphy and M. Robshaw, *Further Comments on the Structure of Rijndael.* August 17, 2000. Archived via http://www.crsc.nist.gov.

27. S. Murphy and M. Robshaw, *Essential algebraic structure within the AES,* Advances in Cryptology – CRYPTO 2002, Lecture Notes in Computer Science, vol. 2442, Springer-Verlag, Berlin – New York, 2002, pp. 1-16.

28. National Institute of Standards and Technology: Advanced encryption standard, FIPS 46-3, US Department of Commerce, Washington D.C., October 1999.

29. National Institute of Standards and Technology: Advanced encryption standard, FIPS 197, US Department of Commerce, Washington D.C., November 2001.

30. S. Park, S.H. Sung, S. Chee, E.-J. Yoon, and J. Lim, *On the security of Rijndael-like structures against differential and linear cryptanalysis,* Advances in Cryptology - ASIACRYPT 2002, Y. Zheng (ed.), Lecture Notes in Computer Science, vol. 2501, Springer-Verlag, Berlin – New York, 2002, pp. 176–191.

31. S. Park, S.H. Sung, S. Lee, and J. Lim, *Improving the upper bound on the maximum differential and the maximum linear hull probability for SPN structures and AES,* Fast Software Encryption, 10th International Workshop, FSE 2003, Lund, Sweden, February 2003, T. Johansson (ed.), Lecture Notes in Computer Science, vol. 2887, Springer-Verlag, Berlin – New York, 2003, pp. 247–260.

32. R.C.W. Phan, *Classes of impossible differentials of the advanced encryption standard,* Electronics Letters 38(11), 2002, pp. 508–510.

33. R.C.W. Phan, *Impossible differential cryptanalysis of 7-round Advanced Encryption Standard,* Information Processing Letters 91, 2004, pp. 33–38.

34. R.C.W. Phan and M.U. Siddiqi, *Generalised impossible differentials of the Advanced Encryption Standard,* Electronics Letters 37(14), 2001, pp. 896–898.

35. H. Raddum, *More Dual Rijndaels,* these proceedings, pp. 144–150.

# The Boomerang Attack on 5 and 6-Round Reduced AES*

Alex Biryukov

Katholieke Universiteit Leuven, Dept. ESAT/SCD-COSIC,
Kasteelpark Arenberg 10,
B–3001 Heverlee, Belgium
http://www.esat.kuleuven.ac.be/~abiryuko/

**Abstract.** In this note we study security of 128-bit key 10-round AES against the boomerang attack. We show attacks on AES reduced to 5 and 6 rounds, much faster than the exhaustive key search and twice faster than the "Square" attack of the AES designers. The attacks are structural and apply to other SPN ciphers with incomplete diffusion.

## 1   Introduction

In this paper we study security of 128-bit key AES [4] against the boomerang attack [7]. The boomerang attack was developed in 1999 after the AES competition was already running. This attack sometimes allows to break more rounds than the conventional differential or linear attacks, especially for the ciphers with few but carefully designed rounds (for example, see an attack on SAFER++ [2]).

In this paper we show attacks on AES reduced to 5 and 6 rounds. Six round attack has complexity of $2^{71}$ data and steps of analysis (measured in 6-round encryptions). The attack is twice faster than the "Square" attack of the designers of the AES in terms of time complexity which is a dominant factor, but has much higher data complexity. The boomerang attack on AES is less efficient than the partial sum attack [5]. See Table 1 for comparison of our attacks with the previous results on a 128-bit key AES.

## 2   Boomerang Attack on SPNs with Incomplete Diffusion

Boomerang attack is a chosen plaintext-adaptive chosen ciphertext attack. It is an extension of differential cryptanalysis and works on quartets of data $(P, P')$,

---

* This work was supported in part by the Concerted Research Action (GOA) Mefisto-2000/06 of the Flemish Government and in part by the European Commission through the IST Programme under Contract IST-2002-507932 ECRYPT. The information in this document reflects only the author's views, is provided as is and no guarantee or warranty is given that the information is fit for any particular purpose. The user thereof uses the information at its sole risk and liability.

H. Dobbertin, V. Rijmen, A. Sowa (Eds.): AES 2004, LNCS 3373, pp. 11–15, 2005.
© Springer-Verlag Berlin Heidelberg 2005

**Table 1.** Comparison of our results with previous attacks on AES

| Attack | Key size | Rounds | Data[a] | Type[b] | Workload[c] | Memory[a] |
|---|---|---|---|---|---|---|
| Square attack [4] | 128 | 5 of 10 | $2^{11}$ | CP | $2^{40}$ | $2^{11}$ |
| Square attack [4] | 128 | 6 of 10 | $2^{32}$ | CP | $2^{72}$ | $2^{32}$ |
| Collision attack [6] | 128 | 7 of 10 | $2^{32}$ | CP | $2^{128}$ | $2^{80}$ |
| Partial sum [5] | 128 | 6 of 10 | $2^{34.6}$ | CP | $2^{44}$ | $2^{32}$ |
| Partial sum [5] | 128 | 7 of 10 | $2^{128}-2^{119}$ | CP | $2^{120}$ | $2^{64}$ |
| Imposs. diff. [1] | 128 | 5 of 10 | $2^{29.5}$ | CP | $2^{31}$ | $2^{40}$ |
| Imposs. diff. [3] | 128 | 6 of 10 | $2^{91.5}$ | CP | $2^{122}$ | $2^{?}$ |
| Our Boomerang attack | 128 | 5 of 10 | $2^{39}$ | CP/ACC | $2^{39}$ | $2^{33}$ |
| Our Boomerang attack | 128 | 6 of 10 | $2^{71}$ | CP/ACC | $2^{71}$ | $2^{33}$ |

[a] Expressed in the number of blocks.
[b] CP – Chosen Plaintext, ACC – Adaptive Chosen Ciphertext.
[c] Expressed in equivalent number of encryptions.

$(Q, Q')$. The attack works when encryption $E()$ can be split into $E = E_1 \circ E_0$, where $E_0$ is weak in encryption direction and $E_1$ is weak in decryption direction. We refer the reader to [7] for further details.

In this section we present a generic method of breaking five and six round substitution-permutation networks (SPNs) using a boomerang distinguisher. The attacks that we will show will be *structural* in the sense that they will not use specific properties of S-boxes or of the mixing layer, but will use only the fact that diffusion is incomplete (which is the case for many ciphers, including the AES).

We will describe this attack on an example of Rijndael-like cipher with layers of 16, 8x8-bit S-boxes, and RIJNDAEL-like diffusion involving ShiftRows and Mix-Columns (though exact constants in the MixColumns matrix will be irrelevant to the attack).

The five round attack will be as follows:

1. Prepare a pool of plaintexts $\{P_i\}, i = 0, \ldots 2^{32} - 1$ which have all possible values in four bytes (which will appear in the same column before the Mix-Columns) and arbitrary constant in the other bytes. Encrypt the pool and obtain a pool of $2^{32}$ ciphertexts $\{C_i\}$.
2. Construct a pool of modified ciphertexts: $D_i = C_i \oplus \nabla$, where $\nabla$ is a fixed non-zero difference with only one active S-box (for example, a non-zero difference in the first byte and zero difference in 15 other bytes).
3. Decrypt the pool $\{D_i\}$ to obtain a pool $\{Q_i\}$ of $2^{32}$ new plaintexts.
4. Sort the pool $\{Q_i\}$ by the bytes corresponding to eight inactive S-boxes. Pick only those pairs $Q_i, Q_j$ which have zero difference in these 8 bytes. If none found go to step 1.
5. For each of the quartets $P_i, P_j, Q_i, Q_j$ that pass step 4, guess the 32-bit key value that enters the four S-boxes corresponding to non-constant bytes. Using the guessed key value partially encrypt one round and check that

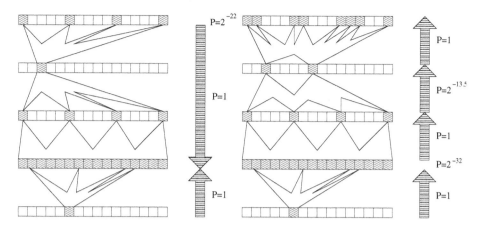

**Fig. 1.** Schematic description of a boomerang quartet for Rijndael reduced to five rounds

resulting difference is in a single active S-box, which is a 22-bit filtering condition for each pair $(P_i, P_j)$ and $(Q_i, Q_j)$. This gives a 44-bit condition in total for both sides of the boomerang in the case of common 4-tuples of active S-boxes. However with probability half we will have no common 4-tuples, i.e. all the twelve active S-boxes (4 from the $(P_i, P_j)$ pair and 8 from the $(Q_i, Q_j)$ pair) non-overlapping. We will then pick key-candidates that are suggested at least twice.

See Figure 1 for a schematic description of the boomerang distinguisher used in this 5-round attack. The rectangles in the figure denote the layers of S-boxes, and the gray squares indicate the active S-boxes. Arrows represent the cost of pattern propagation in terms of probability.

**Complexity of the 5-Round Attack.** Analysis of the complexity of the attack described above is as follows: Each pool of $2^{32}$ texts contains $2^{63}$ pairs with difference in the four relevant bytes. From these pairs $2^{63}/2^{22} = 2^{41}$ will have a single active S-box after one round[1]. After the second round we will have four active S-boxes. After the 3rd round all bytes will be active. From the bottom up direction we will have one round crossed with probability one, with a truncated differential that starts with one active S-box and ends with four active S-boxes. At this point we need that after the next S-box layer the difference in these four active S-boxes would be the same. This happens with probability $2^{-32}$. Then we can switch to the last face of the boomerang, where the effect of the mixing of the 3rd round can then be undone with probability one and we will get four active S-boxes after the S-box layer of the 3rd round. We will have to pay $2^{-13.5}$

---

[1] The chance of going from 4 active S-boxes to one is $4 \cdot 2^8/2^{32} = 2^{-22}$, since we do not care about the location of the single active S-box.

in probability for the four active S-box difference to turn into two active S-box difference after this S-box layer and the MixColumns which is just above it. From that point we let our truncated differential run freely with probability 1. As a result we will obtain a new pair of plaintexts with some difference in eight bytes and zero difference in the other eight bytes. This is our $64 - \log(6)$-bit filtering condition for the good boomerang quartets (the $-\log(6)$ appears since we do not fix the places of the two active S-boxes).

We pick about $2^6$ pools in which we will have $2^{41} \cdot 2^6 \cdot 2^{-13.5} \cdot 2^{-32} \approx 3$ good boomerangs returning back. The average amount of false quartets which satisfy our initial filtering condition is $2^{63} \cdot 2^6 \cdot 2^{-64} \cdot 6 = 192$.

In the simplest case when the boomerang returns in the same four bytes as it was sent we perform a guess of the 32-bit key and check it against two sides of the boomerang $P_i, P_j$ and $Q_i, Q_j$ whether in both cases it leads to a single active S-box after one round. This gives a 44-bit filtration condition which leaves only the correct key guess with probability $1 - 2^{-12}$. However with probability $1/2$ it may happen that for the two boomerang pairs active 4-tuples of the output pair $(Q_i, Q_j)$ will be different from those of the input pair $(P_i, P_j)$. In this case we independently guess 32-bits of the key corresponding to the input four bytes for each pair and leave only those keys that lead to a single active S-box after the first round. We expect at least two good boomerang quartets in our pools and thus the correct key would be counted at least twice. Note that we have about 100 noisy pairs (those with non-overlapping 4-tuples) each of which suggests about $4 \cdot 2^8$ candidates for the 32-bit key. That means that we may have a few wrong keys suggested due to the birthday collisions together with the correct key suggested by the good boomerangs. The same analysis can be performed in parallel on another 4-tuple to produce few candidates for another 32-bit part of the key. Knowing a few candidates for at least half of the first subkey we can repeat the attack with much less data and smaller complexity to achieve the full key-recovery.

Total complexity of this 5-round attack is $2^{38}$ chosen plaintext/adaptive chosen ciphertext queries and $2^{38}$ time steps which mainly would be spent on encrypting and sorting the data. The memory required by the attack is $2^{32}$ blocks or $2^{36}$ bytes.

**Extension to 6-Rounds of AES.** The attack described above can be extended by one round at the bottom at the cost of guessing 32-bits of the key of the 6th round. We double the number of pools from $2^6$ to $2^7$ to get 4-6 good quartets for better filtration. We expect that at least 2-3 good quartets will have overlapping 4-tuples between the $P$'s and the $Q$'s which provides $2^{-12}$ filtration power. Thus we will get about $2^{32} \cdot 100 \cdot 2^{-12} \approx 2^{27}$ candidates for 64-bit partial key: 32-bits at the top and 32-bits at the bottom. The correct key will be suggested at least twice, and the wrong keys would likely be suggested only once, since we are below the birthday bound for a 64-bit event. Thus we expect that all the wrong pairs will be filtered out at the key-recovery step. Finally, complexity of this 6 round attack will be $2^{39}$ chosen plaintexts, $2^{71}$ adaptively chosen ciphertexts, the same amount of time steps spent mainly encrypting the texts and $2^{37}$ bytes of memory.

It seems likely that this attack may be converted to break 7-rounds of the 192-bit key AES.


# 3   Conclusions

We have shown boomerang attacks on 5 and 6 round AES much faster than exhaustive search. We notice that AES has many truncated differentials with probability one spanning up to three rounds, however they are quite expensive in terms of probability when trying to extend them at either end of the boomerang distinguisher. The attacks presented in this paper are twice more efficient than the "Square" attack but are less efficient than the partial sum attack. This may mean that AES has sufficient security margin against the boomerang attacks. The attacks presented in this paper are *structural* attacks (i.e. they do not use specific properties of the underlying cipher) applicable to arbitrary 5-6 round SPNs with incomplete diffusion. It is an open problem whether the middle-round gaining trick, for example as used in a recent attack on Safer++ [2] would be applicable to the AES.


# References

[1] E. Biham and N. Keller, "Cryptanalysis of reduced variants of Rijndael," in *Official public comment for Round 2 of the Advanced Encryption Standard development effort*, 2000. Available at http://csrc.nist.gov/encryption/aes/round2/conf3/papers/35-ebiham.pdf.

[2] A. Biryukov, C. D. Canniére, and G. Dellkrantz, "Cryptanalysis of SAFER++," in *Proceedings of Crypto'03* (D. Boneh, ed.), Lecture Notes in Computer Science, Springer-Verlag, 2003. NES/DOC/KUL/WP5/028. Full version available at http://eprint.iacr.org/2003/109/.

[3] J. H. Cheon, M. Kim, K. Kim, J.-Y. Lee, and S. Kang, "Improved impossible differential cryptanalysis of Rijndael and Crypton," in *Proceedings of ICISC'01* (K. Kim, ed.), no. 2288 in Lecture Notes in Computer Science, pp. 39–49, Springer-Verlag, 2001.

[4] J. Daemen and V. Rijmen, *The Design of Rijndael: AES — The Advanced Encryption Standard*. Springer-Verlag, 2002.

[5] N. Ferguson, J. Kelsey, S. Lucks, B. Schneier, M. Stay, D. Wagner, and D. Whiting, "Improved cryptanalysis of Rijndael," in *Fast Software Encryption, FSE 2000* (B. Schneier, ed.), vol. 1978 of *Lecture Notes in Computer Science*, pp. 213–230, Springer-Verlag, 2001.

[6] H. Gilbert and M. Minier, "A collision attack on seven rounds of Rijndael," in *Proceedings of the Third AES Candidate Conference*, pp. 230–241, National Institute of Standards and Technology, Apr. 2000.

[7] D. Wagner, "The boomerang attack," in *Fast Software Encryption, FSE'99* (L. R. Knudsen, ed.), vol. 1636 of *Lecture Notes in Computer Science*, pp. 156–170, Springer-Verlag, 1999.

# A Three Rounds Property of the AES

Marine Minier

Université Paris 8 - INRIA,
Projet CODES,
Domaine de Voluceau-Rocquencourt,
B.P. 105, 78 153 Le Chesnay Cedex - France
marine.minier@inria.fr

**Abstract.** Rijndael is the new Advanced Encryption Standard designed by V. Rijmen and J. Daemen and chosen as AES by the NIST in October 2000. Surprisingly, the number of cryptanalyses against this algorithm is very low in depict of many efforts furnished to break it.

This paper presents a stronger property than the one used in the Bottleneck Cryptanalysis [GM00]. Unfortunately, this property could not be used to mount a more efficient cryptanalysis than the Bottleneck Attack because it is not possible to improve the complexity of the four rounds distinguisher used in this attack. So, the complexity of the Bottleneck Attack (recalled in this paper) is always $2^{144}$ AES executions using $2^{32}$ plaintexts.

## 1  Introduction

In the initial article describing Rijndael [DR98], V. Rijmen and J. Daemen wrote : "For the different block lengths of Rijndael, no extensions to 7 rounds [of a known attack] faster than an exhaustive key search have been found". Of course, since 1998, some attacks reached this aim. In the case of key length equal to 192 or 256 bits, Ferguson et al., in [FKS+00], presented an improvement of the Square Attack [DR98] permitting to cryptanalyse an eight-rounds version of Rijndael with a complexity equal to $2^{204}$ executions and $2^{128} - 2^{119}$ plaintexts. S. Lucks presented in [Luc01] an other improvement of the Square Attack using a particular weakness of the key schedule against a seven-rounds version of Rijndael where $2^{194}$ executions are required for a number of chosen plaintexts equal to $2^{32}$. H. Gilbert and M. Minier in [GM00] also presented an attack against a seven-rounds version of Rijndael (known under the name of "Bottleneck Attack") using a stronger property on three inner rounds than the one used in the Square Attack in order to mount an attack against a seven-rounds version of Rijndael requiring $2^{144}$ cipher executions with $2^{32}$ chosen plaintexts. In the case of a 128 bits key lenght, for a seven-rounds version of Rijndael, only two attacks are known. The first is due to Ferguson et al. in [FKS+00] and requires $2^{120}$ cipher executions for a number of plaintexts equal to $2^{128} - 2^{119}$. The second one, due to H. Gilbert and M. Minier in [GM00], is a marginal speed up of the 128-bits key search requiring $2^{32}$ chosen plaintexts.

H. Dobbertin, V. Rijmen, A. Sowa (Eds.): AES 2004, LNCS 3373, pp. 16–26, 2005.

During the two last years, some new results was published concerning essentially the algebraic structure of the AES S-box. Those results use a potential weakness of the AES : there is only one non-linear operation in the AES round function, the inversion in the Galois Field $GF(2^8)$. So, it is possible to derive this inversion application into quadratic equations that are true with probability one. In [CP02], N. Courtois and J. Pieprzyk presents the quadratic equations given by the AES S-box on $GF(2)$. The authors use those quadratic equations to express all input/output bytes of each round and generate a huge system to solve. They apply the XL and the XSL algorithms to obtain the solutions of the system generated. An other article presented at Crypto'02 by S. Murphy and M. Robshaw [MR02] also describe the algebraic structure of the AES S-box and the quadratic equations of this one but on the field $GF(256)$.

The aim of this paper is to present a "new property" on three inner AES rounds stronger than the one used in the Bottleneck Attack described in [GM00]. This "new property" is very similar to the bottleneck property but as now does not permit to improve any attack due to the same number of dependent bytes implied in the four rounds distinguisher. This property is, however, stronger because the number of deduced collisions is bigger.

This paper is organized as follow: Section 2 provides a brief outline of the AES. Section 3 describes the 3-rounds property and the 4-rounds distinguisher used in the Bottleneck Attack. Section 4 presents the "new property". Section 5 describes, one more time, the bottleneck attack on seven rounds of the AES for a 128 bits block and a 192 or 256 bits key. Section 6 concludes this paper.

## 2    A Brief Outline of the AES

The AES is a symmetric block cipher using a parallel and byte-oriented structure. The key length and the block length are variable and are equal to 128, 192 or 256 bits. The current block is represented by a matrix of bytes. We focus from now on a 128-bits block represented by a $4 \times 4$ matrix of bytes :

$$B = \begin{array}{|c|c|c|c|} \hline b_{0,0} & b_{0,1} & b_{0,2} & b_{0,3} \\ \hline b_{1,0} & b_{1,1} & b_{1,2} & b_{1,3} \\ \hline b_{2,0} & b_{2,1} & b_{2,2} & b_{2,3} \\ \hline b_{3,0} & b_{3,1} & b_{3,2} & b_{3,3} \\ \hline \end{array}$$

The number of rounds $nr$ is also variable : 10, 12 or 14, depending on the block length and on the key length. The key schedule derives $nr + 1$ 128-bits round keys $k_0$ to $k_{nr}$ from the master key $k$ of variable length.

The round function, repeated $nr - 1$ times, is composed of four basic transformations, all linear except the first one :

- SubBytes : a bytewise transformation that applies on each byte of the current block an 8-bits to 8-bits non linear S-box (that we call $S$) composed by the inversion in the Galois Field $GF(256)$ and by an affine transformation.

- ShiftRows: a linear mapping that rotates on the left all the rows of the current matrix (0 for the first row, 1 for the second, 2 for the third and 3 for the fourth)
- MixColumn: another linear mapping represented by a $4 \times 4$ matrix chosen for its good properties of diffusion (see [DR02]). Each column of the input matrix is multiplied by the MixColumns matrix in the Galois Field $GF(256)$ that provides the corresponding column of the output matrix. We denote by $a_{i,j}$ for $i$ and $j$ from 0 to 3, the coefficients of the MixColumns matrix.
- AddRoundKey : a simple x-or operation between the input matrix and the subkey of the current round denoted by $k_i$.

Those $nr - 1$ rounds are surrounded at the top by an initial key addition with the subkey $k_0$ and at the bottom by a final transformation composed by a call to a round function where the MixColumns operation is omitted.

## 3    The Three-Rounds Property and the Four-Rounds Distinguisher

We now describe the three inner rounds property used in [GM00] and the four inner rounds distinguisher deduced.

### 3.1    The Three-Rounds Property

We note Y, Z, R and S the different intermediate input/output states of three consecutive inner rounds as noticed in figure 1.

We focus from now on an input block Y with its three left columns fixed. The most at right column, marked on figure 1, is composed by one active byte $y$ which takes all possible values between 0 and 255 and by a triplet $c$ equal to $(c_0, c_1, c_2)$ of constant bytes which will represent a parameter. More formally, we note $Y_{0,3} = y$, $Y_{1,3} = c_0$, $Y_{2,3} = c_1$ and $Y_{3,3} = c_2$.

In the same way, we use the following notations for some particular bytes marked on figure 1. So, we denote $Z_{0,3} = z_0$, $Z_{1,3} = z_1$, $Z_{2,3} = z_2$, $Z_{3,3} = z_3$, $R_{0,3} = r_0$, $R_{1,0} = r_1$, $R_{2,1} = r_2$, $R_{3,2} = r_3$ and $S_{0,3} = s$.

So, let us analyze how the Z, R and S particular bytes $z_0$ to $z_3$, $r_0$ to $r_3$ and $s$ can be seen as $c$-dependent and key-dependent functions of the $y$ input byte.

- After the first round, the $y \rightarrow z_0^c[y]$ one to one function is independent from the value of the $c$ triplet and is entirely determined by one key byte, due to the effect of the ShiftRows operation. The same property holds for $z_1$, $z_2$ and $z_3$. So, the quartet of bytes $(z_0, z_1, z_2, z_3)$ is a function of the $y$ values entirely determined by four key-dependent bytes. More formally, there exists 4 key-dependent constants $k_{1,i,0}$ for $i = 0..3$ such that

$$z_i = a_{i,0} \cdot S(y) + k_{1,i,0} \ , \ i = 0..3$$

where $S$ represents the AES S-box.

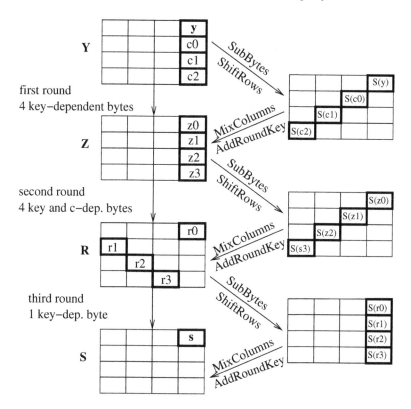

**Fig. 1.** The Three Inner Rounds Property

- After the second round, each of the four bytes $r_i[y]$, $i = 0..3$ is a one to one function of the corresponding $z_i[y]$ byte entirely determined by one single unknown byte that is entirely determined by $c$ and the key. The quartet of bytes $(r_0, r_1, r_2, r_3)$ of R marked on figure 1 is a function of $(z_0, z_1, z_2, z_3)$ entirely determined by four key-dependent and $c$-dependent bytes.
- After the third round, the $s$ byte marked on figure 1 can be expressed as a function of the $(r_0, r_1, r_2, r_3)$ quartet of bytes entirely determined by one key-dependent byte depending on the subkey of the round.

In summary, the $s$ byte depends on only 5 key-dependent bytes and 4 $c$-dependent bytes. More formally, the partial function $s^c[y]$ is entirely determined by a reduced number of unknown bytes. We can exploit this restricted dependency by constructing collisions on all the $y$ values for distinct values of the $c$ triplet. In other words :

**Property 1.** *There exists $c'$ and $c''$ two triplets of constants such as for all $y$ values between 0 and 255, we have : $s^{c'}[y] = s^{c''}[y]$. In this case, we say that we have a collision.*

In fact, the number of obtained collisions is 256, one for each $y$ value.

Under the heuristic assumption that the unknown constants depending on the key and on the $c$ triplet behave as random functions, then, by the birthday paradox, if we take a $C$ set of $2^{16}$ $c$ triplets, the probability to obtain a collision is non negligible.

This property can be extended to mount an efficient four-rounds distinguisher by adding a fourth round at the bottom of the three previous rounds.

## 3.2    The Four-Rounds Distinguisher

We consider the deciphering of the fourth round in the following way (see figure 2) :

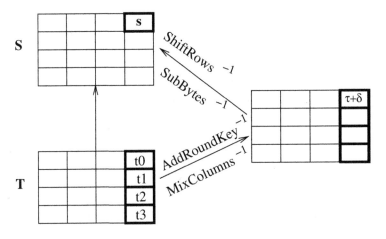

**Fig. 2.** The Extension to a Fourth Round

We denote by T the output block after the fourth round and we denote by $(t_0, t_1, t_2, t_3)$ the last column of T marked on the figure 2. We can express the byte $s$ as $s = S^{-1}[(0E \cdot t_0 + 0B \cdot t_1 + 0D \cdot t_2 + 09 \cdot t_3) + \delta]$ where $S$ represents the AES S-box and $\delta$ a constant depending on the subkey of the fourth round. We have the following property :

**Property 2.** *There exists a collision between $s^{c'}[y]$ and $s^{c''}[y]$ if and only if for all $y$ values between 0 and 255, we have :*

$$0E \cdot t_0^{c'} + 0B \cdot t_1^{c'} + 0D \cdot t_2^{c'} + 09 \cdot t_3^{c'} = 0E \cdot t_0^{c''} + 0B \cdot t_1^{c''} + 0D \cdot t_2^{c''} + 09 \cdot t_3^{c''}.$$

To simplify the notations we denote $0E \cdot t_0^c + 0B \cdot t_1^c + 0D \cdot t_2^c + 09 \cdot t_3^c$ by $\tau^c[y]$. $\tau^c[y]$ is a function of $y$ entirely determined by 6 unknown bytes depending on the key and by 4 additional unknown bytes depending on both the key and the $c$ values.

The following four inner rounds distinguisher is tested on a limited number of $y$ values, a set $\Lambda$ of 16 values is sufficient, the number of false alarms being negligible in this case.

- Select a $C$ set of about $2^{16}$ $c$ triplet values and a subset of $\{0 \cdots 255\}$, say for instance a $\Lambda$ subset of 16 $y$ values.
- For each $c$ triplet value, compute the $L_c = (0E \cdot t_0^c + 0B \cdot t_1^c + 0D \cdot t_2^c + 09 \cdot t_3^c)_{y \in \Lambda}$.
- Check wether two of the above lists, $L_{c'}$ and $L_{c''}$ are equal.

The computations made at the secund step of this distinguisher (16 linear combinations of the outputs) represent substantially less one single AES execution.

This four-rounds distinguisher requires about $2^{20}$ chosen inputs Y, and since the collision detection computations (based on the analysis of the corresponding T values) require less operations than the $2^{20}$ 4-inner rounds computations, the complexity of the distinguisher is less than $2^{20}$ AES encryptions for a probability of success equal to $1/2$ (due to the birthday paradox).

# 4    The "New Three-Rounds Property"

We describe, in this section, a "new" three-rounds property derived from the previous one that permits to obtain an higher number of collisions. For more clarity, we use the same notation than the previous one. In this "new property", the number of initial active bytes in the last Y column has been modified. Here, we define two active bytes $y$ and $c_0$ instead of one (the $y$ byte) before. In this case, the $c$ triplet becomes a pair of bytes defined by $c_p = (c_1, c_2)$. Let us explain how those two active bytes cross three inner rounds (see figure 3) :

- After the first round, the $y \to z_0^{c_p}[y, c_0]$ one to one function is independent from the value of the $c_p$ triplet and is entirely determined by one key byte, due to the effect of the ShiftRows operation. The same property holds for $z_1$, $z_2$ and $z_3$. So, the quartet of bytes $(z_0, z_1, z_2, z_3)$ is a function of the $y$ and $c_0$ values entirely determined by four key-dependent bytes.
- After the second round, each of the four bytes $r_i[y, c_0]$, $i = 0..3$ is a one to one function of the corresponding $z_i[y, c_0]$ byte entirely determined by one single unknown byte that is entirely determined by $c_p$ and the key. The quartet of bytes $(r_0, r_1, r_2, r_3)$ of R marked on figure 3 is a function of $(z_0, z_1, z_2, z_3)$ entirely determined by four key-dependent and $c$-dependent bytes.
- After the third round, the $s$ byte marked on figure 3 can be expressed as a function of the $(r_0, r_1, r_2, r_3)$ quartet of bytes entirely determined by one key-dependent byte depending on the subkey of the round.

As in the section 3.1, we can deduce that the $s$ byte at the end of the third round is a function of $y$ and $c_0$ entirely determined by 5 key-dependent bytes and 4 key-dependent and $c_p$-dependent bytes. We can also exploit this restricted dependency between the $s$ byte and the two active bytes $y$ and $c_0$ by defining a new kind of collision :

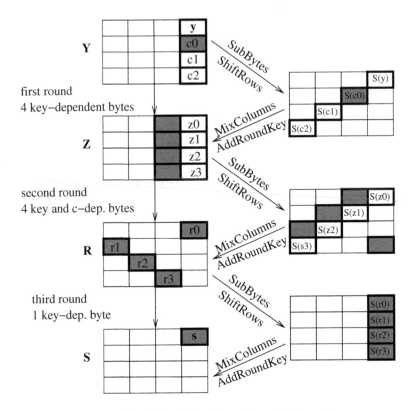

**Fig. 3.** The Other Three Inner Rounds Property

**Property 3.** *There exists $c'_p$ and $c''_p$ two pairs of constants such as for all $y$ and $c_0$ values between 0 and 255, we have : $s^{c'_p}[y, c_0] = s^{c''_p}[y, c_0]$. In this case, we say that we have a collision.*

The number of obtained collisions is $(256)^2$, one for each $y$ and $c_0$ value.

This "new" property is due to the very symmetric and parallel structure of the AES in the byte position level.

We verify the veracity of this property by computer experiments.

Unfortunately, this property could not be used to mount a more efficient four-rounds distinguisher than the one presented in section 3.2. Indeed, the number of $c_p$ pairs used in the distinguisher and the probability of success depend on only the four intermediate $c_p$-dependent bytes. The number of such bytes is the same for the property of the section 3.1 and the "new" one in depict of the bigger number of obtained collisions.

So, we do not find a more efficient distinguisher that permits to use this stronger property and to improve an attack. We use the same distinguisher than the one described in section 3.2.

# 5    The Bottleneck Attack on a Seven-Rounds Version of the AES with Key Lengths Equal to 192 and 256 Bits Using $2^{32}$ Chosen Plaintexts

Even if the new property could not be used to improve the four-rounds distinguisher described in section 3.2 and so the bottleneck attack, we give a short description of this known cryptanalysis. So, we are going to recall, in this section, how the four inner rounds distinguisher of the section 3.2 could be extended to mount a seven-rounds attack on the AES with a 192 or a 256 bits key.

The seven-rounds version of the AES is depicted in figure 4. The seventh round is here considered as the last round (i.e. it doesn't contain the Mix-Columns operation). X represents a plaintext block and V the corresponding ciphertext. The four previous rounds are surrounded at the top by one initial round X $\rightarrow$ Y, composed by an initial key addition followed by one round and at the bottom by two final rounds : T $\rightarrow$ U and U $\rightarrow$ V.

The attack method is a combination between the four-rounds distinguisher presented in section 3.2 and an exhaustive search of some keybytes or combination of keybytes of the initial and the two final rounds. The attack described here uses the fact that, in the equations provided by the four-rounds distinguisher, there is a variables separation in terms which involve one half of the 2 last rounds key bytes and terms which involve a second half of the 2 last round key bytes in order to save a $2^{80}$ factor in the exhaustive search complexity.

## 5.1    Extension at the Beginning

The distinguisher of section 3.2 could be extended by one round at the beginning using the same method than the one proposed by the authors of Rijndael in the initial paper [DR98] and first applied to the algorithm Square.

The main idea used here is that if, in the initial key addition, the 4 key bytes (denoted by $k_{ini} = (k_{0,0}, k_{0,1}, k_{0,2}, k_{0,3})$), added with the four bytes $(x_0, x_1, x_2, x_3)$ of the plaintext X marked in figure 3, are known then it is possible to partition the $2^{32}$ plaintexts into $2^{24}$ subsets of $2^8$ plaintexts values satisfying the conditions of section 3.2 (i.e. $(c_0, c_1, c_2)$ stay a triplet of constants and $y$ is the active byte).

So, if all the $2^{32}$ possible plaintexts are encrypted for all the possible values of the $(x_0, x_1, x_2, x_3)$ quartet (the other 12 bytes being taken equal to a constant), the $2^{32}$ plaintexts could be partitioned, according to the value of $k_{ini}$, into $2^{24}$ subsets of $2^8$ plaintexts according the values of $y$ (which are known up to an unknown constant linked with the first round key byte). Those subsets are such that the $y$ byte takes all possible values between 0 and 255 and the $c = (c_0, c_1, c_2)$ triplet is composed of three constant values, different and unique for each of the $2^{24}$ subsets, the 12 other Y bytes are constant and all those constant values are the same for all subsets.

Those $2^{32}$ plaintexts give the corresponding $2^{32}$ ciphertexts $V$.

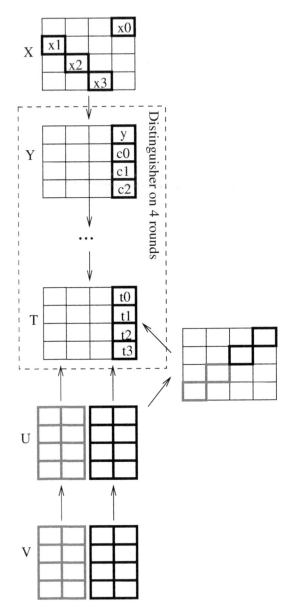

**Fig. 4.** Attack on Seven Rounds of the AES

## 5.2  Extension at the End

Each of the $t_0, t_1, t_2, t_3$ bytes can be expressed as a function of four bytes of the V ciphertext and five unknown key bytes (i.e. four of the final round subkey and one linear combination of the penultimate round subkey). So, in order to improve the key exhaustive search on the two last rounds, the equations of collisions are "cuted" in two parts as follows :

$$\tau_1^{c'} = 0E \cdot t_0^{c'} + 0B \cdot t_1^{c'} \text{ and } \tau_2^{c'} = 0D \cdot t_2^{c'} + 09 \cdot t_3^{c'}$$

With this notation the equation of collision $\tau^{c'} = \tau^{c''}$ described in property 2 could be expressed as $\tau_1^{c'} + \tau_2^{c'} = \tau_1^{c''} + \tau_2^{c''}$, i.e. $\tau_1^{c'} + \tau_1^{c''} = \tau_2^{c'} + \tau_2^{c''}$. $\tau_1$ depends on $t_0$ and $t_1$ and $\tau_2$ depends on $t_2$ and $t_3$. Now, due to the fact that the last round of the AES does not contain the MixColumns operation, $t_0$ and $t_1$ could be expressed, as shown on figure 3, as a function of 8 ciphertext bytes and 10 key bytes of the two last rounds denoted by $k_{\tau_1}$. In the same way, $t_2$ and $t_3$ depend on 8 ciphertext bytes and on 10 key bytes of the two last rounds denoted by $k_{\tau_2}$.

This remark permits to share in two parts the key exhaustive search and to improve the attack on a seven rounds-version of the AES by a factor $2^{80}$.

### 5.3   Outline of the Attack

An efficient exhaustive search of the $k_{ini}$, $k_{\tau_1}$ and $k_{\tau_2}$ keys could be performed in the following way:

**First step :**
    Cipher the $2^{32}$ chosen plaintexts for all possible values of the quartet $(x_0, x_1, x_2, x_3)$.
**Second step :**
For $k_{ini}$ from $(0,0,0,0)$ to $(255,255,255,255)$ do
        Partition the $(256)^4$ chosen plaintexts
        into $(256)^3$ $\Lambda^c$ sets according the value of the triplet $c$
        Choose into those $(256)^3$ $\Lambda^c$ sets $2^{16}$ values of $c$
        For each value of the $(c', c'')$ pair do
            For $k_{\tau_1}$ from $(0, \cdots, 0)$ to $(255, \cdots, 255)$ do
                Compute the values of $(\tau_1^{c'} \oplus \tau_1^{c''})_{y=0\cdots 15}$ from the ciphertexts
                Put them in a table $T_{k_{ini},c',c''}[k_{\tau_1}]$
            End For
            For $k_{\tau_2}$ from $(0, \cdots, 0)$ to $(255, \cdots, 255)$ do
                Compute the values of $(\tau_2^{c'} \oplus \tau_2^{c''})_{y=0\cdots 15}$ from the ciphertexts
                Look in the table $T_{k_{ini},c',c''}[k_{\tau_1}]$ if the same values appear
                If yes, verify the same computation for all the $y$ values
                    If equality for all $y$ values, return $(k_{ini}, k_{\tau_1}, k_{\tau_2})$
                Else continue
                End If
            End For
        End For
    End For
End For

Since the above procedure tests whether the exist collisions inside a random set of $256^2$ of the $256^4$ possible $s^c[y]$ functions, the probability of the procedure to result in a collision, and thus to provide $k_{ini}$, $k_{\tau_1}$ and $k_{\tau_2}$ is high (say about $1/2$). In other words, the success probability of the attack is about $1/2$.

The first step could be made independently and requieres $2^{32}$ chosen plaintexts and $2^{32}$ AES executions.

The complexity of the secund step is about $2^{144}$ operations less expensive than AES executions. Its probability of success is about $1/2$. This attack provides 20 bytes of information on the last and penultimate key values.

### 5.4    How to Improve this Attack Using the Lucks' Property of the Key Schedule for a 192 Bits Key

We can improve, by using the particular property of the key schedule described by S. Lucks in [Luc00], the complexity of the attack by a little factor in the case of a key length equal to 192 bits. Indeed, the attack presented by S. Lucks permits to limit the key exhaustive search to only 8 $k_{\tau_2}$ bytes instead of the 10 initial bytes because the knowledge of the two first columns of the last subkey determines completely, taking into account the effect of the last ShiftRow, the two others bytes of the penultimate subkey that compose $k_{\tau_2}$.

## 6    Conclusion

We have shown in this paper that there exists a strong collision property on three inner AES rounds due to some partial byte oriented functions induced by the AES cipher. This property is stronger than the one used in the bottleneck attack even if this new bottleneck property could not be extend in a better four rounds distinguisher that the one used in the known attack.

Maybe, there is a better way to exploit this new restricted dependency but we do not find how to extend it.

## References

[AES99]    http://www.nist.gov/aes
[CP02]    N. Courtois and J. Pieprzyk, "Cryptanalysis of Block Ciphers with Overdefined Systems of Equations". In *Asiacrypt'02*, Queenstown, New Zealand, Lecture Notes in Computer Science 2501, Springer-Verlag, 2002.
[DR98]    J. Daemen, V. Rijmen, "AES Proposal : Rijndael", *The First Advanced Encryption Standard Candidate Conference,* N.I.S.T., 1998.
[DR02]    J. Daemen, V. Rijmen, *The Design of Rijndael.* Springer-Verlag, 2002.
[FKS+00]    N. Ferguson, J. Kelsey, B. Shneier, M. Stay, D. Wagner and D. Whiting, "Improved Cryptanalysis of Rijndael". In *Fast Software Encryption'00*, New York, United State, pp. 213-230. Lectures Notes in Computer Science 1978, Springer-Verlag, 2000.
[GM00]    H. Gilbert, M. Minier, "A Collision Attack on 7 rounds of Rijndael". In *The Third Advanced Encryption Standard Candidate Conference.* N.I.S.T., 2000.
[Luc00]    S. Lucks, "Attackng Seven Rounds of Rijndael Under 192-bit and 256-bit Keys". In *The Third Advanced Encryption Standard Candidate Conference.* N.I.S.T., 2000.
[MR02]    S. Murphy and M. Robshaw, "Essential Algebraic Structure Within the AES". In *Crypto'02*, Santa Barbara, United State, Lectures Note in Computer Science 2442, Springer-Verlag, 2002.

# DFA on AES

Christophe Giraud

Oberthur Card Systems,
25, rue Auguste Blanche, 92 800 Puteaux, France
c.giraud@oberthurcs.com

**Abstract.** In this paper we describe two different DFA attacks on the AES. The first one uses a fault model that induces a fault on only one bit of an intermediate result, hence allowing us to obtain the key by using 50 faulty ciphertexts for an AES-128. The second attack uses a more realistic fault model: we assume that we may induce a fault on a whole byte. For an AES-128, this second attack provides the key by using less than 250 faulty ciphertexts.

If we extend our hypothesis by supposing that the attacker can choose the byte affected by the fault, our bit-fault attack requires 35 faulty ciphertexts to obtain the secret key and our byte-fault attack requires only 31 faulty ciphertexts.

**Keywords:** AES, DFA, side-channel attacks, smartcards.

## 1 Introduction

Since Boneh, Demillo and Lipton introduced a cryptanalytic attack in September 1996 based on the fact that errors may be induced on smartcards during the computation of a cryptographic algorithm to find the key [6], many papers have been published on this subject. Boneh *et al.* succeeded in breaking an RSA CRT with both a correct and a faulty signature of the same message. Lenstra then improved their attack [9] by finding one of the factors of the public modulus using only one faulty signature of a known message. In October 1996, Biham and Shamir published an attack on secret key cryptosystems [4] entitled Differential Fault Analysis (DFA). In 2000, Biehl, Meyer and Müller presented a paper describing two types of DFA attacks on elliptic curve cryptosystems [3] which were later refined by Ciet and Joye [7].

DFA is frequently used nowadays to test the security of cryptographic smartcards applications, especially those using the DES. On the 2nd October 2000, the AES was chosen to be the successor of the DES and, since then, it is used more and more in smartcards applications. So it seems interesting to investigate what is feasible on the AES by using DFA. Unfortunately, the DFA attack on symmetric cryptosystems proposed by Biham and Shamir [4] does not work on the AES. This is why we work to find a way to attack the AES by using DFA.

On a smartcard, a fault may be induced by its owner in many ways, such as power glitch, clock pulse or radiation of many kinds (laser, etc...). These external

H. Dobbertin, V. Rijmen, A. Sowa (Eds.): AES 2004, LNCS 3373, pp. 27–41, 2005.

interventions may induce a fault, but we do not know the real impact on the computation inside the card. This is why, in this paper, we use two types of fault models. The first fault model assumes that the fault occurs on only one bit of a temporary result. Of course such a fault may be difficult to induce in practice, so the second fault model assumes that the induced fault may change a whole byte. The first fault model is the same as the one used in [3, 4, 6] and was put into practice in 2002 by Skorobogatov and Anderson [13].

In the course of this paper, we describe the AES algorithm before looking at a DFA attack on the AES by using our first fault model. This attack allows us to find the AES-128 key by using 50 faulty ciphertexts. We then explain a more practical DFA attack on an AES-128 by using our second fault model. This attack allows us to find the key by using less than 250 faulty ciphertexts. Finally we present the second attack on a real smart card from a practical point of view.

## 2    AES

In the rest of the paper, we will use the following notations:

- we denote by $M$ the plaintext and by $K$ the AES key,
- $M^i$ denotes the temporary cipher result after the $i^{\text{th}}$ round and $M^i_j$ the $j^{\text{th}}$ byte of $M^i$,
- $K^i$ denotes the $i^{\text{th}}$ AES round key and $K^i_j$ the $j^{\text{th}}$ byte of $K^i$,
- $C$ denotes the correct ciphertext and $C_j$ the $j^{\text{th}}$ byte of $C$,
- $D$ denotes a faulty ciphertext and $D_j$ the $j^{\text{th}}$ byte of $D$.

The following section gives a general description of the AES. For more information, the reader can refer to [11, 8].

### 2.1    General Description

The AES algorithm is capable of encrypting or decrypting data blocks of 128 bits by using cryptographic keys of 128, 192 or 256 bits.

The AES key scheduling provides $N_r + 1$ round keys. The number of rounds $N_r$ is dependent on the key length as shown in the following table:

|         | Key length | Number of Rounds |
|---------|------------|------------------|
| AES-128 | 128        | 10               |
| AES-192 | 192        | 12               |
| AES-256 | 256        | 14               |

A 16-byte temporary result is represented as a two-dimensional array of bytes consisting of 4 rows and 4 columns. For example, $M^i = (M^i_0, ..., M^i_{15})$ is represented by the following array:

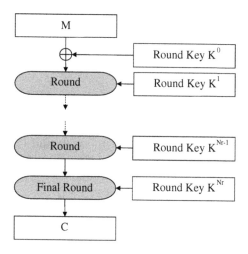

**Fig. 1.** General structure of AES

| $M_0^i$ | $M_4^i$ | $M_8^i$ | $M_{12}^i$ |
|---|---|---|---|
| $M_1^i$ | $M_5^i$ | $M_9^i$ | $M_{13}^i$ |
| $M_2^i$ | $M_6^i$ | $M_{10}^i$ | $M_{14}^i$ |
| $M_3^i$ | $M_7^i$ | $M_{11}^i$ | $M_{15}^i$ |

## 2.2   A Round

The Round function is composed of 4 transformations: SubBytes (SB), ShiftRows (SR), MixColumns (MC) and a bit-per-bit XOR with a round key. The Final Round of the AES is composed of the same functions as a classical Round except that it does not include the MixColumns transformation.

**SubBytes.** This transformation is a non-linear byte substitution and operates on each input byte independently. So, we apply the substitution table (S-box) on each byte of the input to obtain the output.

**ShiftRows.** The rows of the temporary result are cyclically shifted over different offsets. Row 0 is not shifted, Row 1 is shifted over 1 byte, Row 2 is shifted over 2 bytes and Row 3 is shifted over 3 bytes.

**MixColumns.** Here, the columns of the temporary result are considered as polynomials over $\mathbb{F}_{2^8}$ and multiplied modulo $x^4 + 1$ with a fixed polynomial $a(x) = 03 * x^3 + 01 * x^2 + 01 * x + 02$.

Notice that if we change a byte of the input of SubBytes or of ShiftRows, it will change one byte of the output. But for the MixColumns transformation, changing a byte of the input induces a modification of four output bytes.

## 2.3    Key Scheduling

The Key Scheduling generates the round keys from the AES key $K$ by using 2 functions: the Key Expansion and the Round Key Selection.

**Key Expansion.** This function computes from the AES key, an expanded key of length equal to the message block length multiplied by the number of rounds plus 1.

The expanded key is a linear array of 4-byte words and is denoted by $EK[4 * (N_r + 1)]$ where $N_r$ is the number of rounds. If we denote by $N_k$ the key length in words, the key expansion is described in the following pseudo code:

KeyExpansion(byte Key[4 * $N_k$], word EK[4 * ($N_r + 1$)])

```
{
  word temp;
  for (i = 0 ; i < Nk ; i++)
    EK[i] = (Key[4 * i], Key[4 * i + 1], Key[4 * i + 2], Key[4 * i + 3]);
  for (i = Nk ; i < 4 * (Nr + 1) ; i++)
    temp = EK[i − 1];
    if (i mod Nk = 0)
      temp = SubWord(RotWord(temp)) ⊕ Rcon[i/Nk];
    else if ((Nk > 6) and (i mod Nk = 4))
      temp = SubWord(temp);
    EK[i] = EK[i − Nk] ⊕ temp;
}
```

where:

- *SubWord()* is a function that applies the AES S-box at each byte of the 4-byte input to produce an output word,
- *RotWord()* is a cyclic rotation such that a 4-byte input $(a, b, c, d)$ produces the 4-byte output $(b, c, d, a)$,
- the round constant word array, $Rcon[i]$, is defined by $Rcon[i] = (x^{i-1}, \{00\}, \{00\}, \{00\})$ with $x^{i-1}$ being powers of $x$ ($x$ is denoted as $\{02\}$) in the field $\mathbb{F}_{2^8}$.

**Round Key Selection.** This routine extracts the 128-bit round keys from the Expanded Key.

**Example of Key Scheduling for an AES-128.**

| AES Key: | $K$ | | | | | | | |
|---|---|---|---|---|---|---|---|---|
| Expanded Key: | $EK_0$ | $EK_1$ | $EK_2$ | $EK_3$ | $EK_4$ | $EK_5$ | $EK_6$ | $EK_7$ | ... |
| Round Keys: | Round Key 0 | | | | Round Key 1 | | | | ... |

where

- $(EK_0, ..., EK_3)$ is the 128-bit AES key $K$,
- $EK_4 = EK_0 \oplus SubWord(RotWord(EK_3)) \oplus Rcon[1]$,
- $EK_5 = EK_1 \oplus EK_4$,
- $EK_6 = EK_2 \oplus EK_5$,
- $EK_7 = EK_3 \oplus EK_6$, ...

## 3    Bit-Fault Attack

In this section, by using a DFA attack where a fault occurs on only one bit of the temporary cipher result at the beginning of the Final Round, we show how to obtain the entire last round key, i.e. the AES key for an AES-128. For more information about this fault model, the reader can refer to [13].

For the sake of simplicity, we describe the attack on an AES using a 128-bit key.

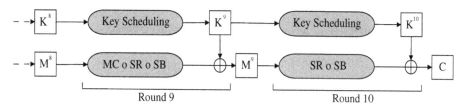

**Fig. 2.** The last rounds of an AES-128

By definition, we have

$$C = ShiftRows(SubBytes(M^9)) \oplus K^{10} \tag{1}$$

Let us denote by $SubByte(M_j^i)$ the result of the substitution table applied on the byte $M_j^i$ and by $ShiftRow(j)$ the position of the $j^{\text{th}}$ byte of a temporary result after applying the $ShiftRows$ transformation.

So, we have from (1)

$$C_{ShiftRow(i)} = SubByte(M_i^9) \oplus K_{ShiftRow(i)}^{10}, \quad \forall i \in \{0, ..., 15\} \tag{2}$$

If we induce a fault $e_j$ on one bit of the $j^{\text{th}}$ byte of the temporary cipher result $M^9$ just before the Final Round, we obtain a faulty ciphertext $D$ where:

$$D_{ShiftRow(j)} = SubByte(M_j^9 \oplus e_j) \oplus K_{ShiftRow(j)}^{10} \tag{3}$$

and for all $i \in \{0, ..., 15\} \backslash \{j\}$, we have:

$$D_{ShiftRow(i)} = SubByte(M_i^9) \oplus K_{ShiftRow(i)}^{10} \tag{4}$$

So, if there is no induced fault on the $i^{\text{th}}$ byte of $M^9$, we obtain from (2) and (4)

$$C_{ShiftRow(i)} \oplus D_{ShiftRow(i)} = 0 \tag{5}$$

and if there is an induced fault on $M_j^9$, we have from (2) and (3)

$$C_{ShiftRow(j)} \oplus D_{ShiftRow(j)} = SubByte(M_j^9) \oplus SubByte(M_j^9 \oplus e_j) \quad (6)$$

Firstly, we determine $ShiftRow(j)$ which is the position of the only non-zero byte of $C \oplus D$ and we thus obtain $j$. We then use a counting method in order to find $M_j^9$: we guess the single bit fault $e_j$ and we find a set of possible values for $M_j^9$ which verify (6). For each of these values, we increase the corresponding counter by 1. With another faulty ciphertext, the right value for $M_j^9$ is expected to be counted more frequently than any wrong value, and can thus be identified. Then we iterate the previous process to obtain all the other bytes of $M^9$.

Now, as we know the value of the ciphertext $C$ and the value of $M^9$, we can easily obtain the last round key $K^{10}$ from the formula (1) and consequently the AES key $K$ by applying the inverse of the Key Scheduling to $K^{10}$.

By using 3 faulty ciphertexts with faults induced on the same byte of $M^9$, we have a 97% chance of having one value left for this byte (cf. appendice A). So, it is possible to obtain the 128-bit AES key by using less than 50 faulty ciphertexts.

This attack operates independently on each byte, so if we succeed in inducing a fault on only one bit on several bytes of $M^9$, we reduce the number of faulty ciphertexts required to obtain the key.

We notice that this attack also operates on the AES-192 and on the AES-256. In such cases, we obtain the last round key, i.e. the security of the AES-192 is reduced from 24 to 8 bytes and the security of the AES-256 is reduced from 32 to 16 bytes.

This attack is powerful but requires inducing a fault on only one bit at the time of a precise event (i.e. at the beginning of the last round) which may be difficult in practice.

## 4 A Second Type of DFA Attack on the AES-128

This DFA attack uses the fault model based on inducing a fault which may change a whole byte of a temporary result. This attack, which only works on an AES using a 128-bit key, is divided into 3 steps :

1. we obtain the last 4 bytes of $K^9$ by exploiting the faulty ciphertexts obtained when a fault is introduced on $K^9$, just before the computation of $K^{10}$,
2. we obtain another 4 bytes of $K^9$ by exploiting the faulty ciphertexts obtained when a fault is introduced on $K^8$, just before the computation of $K^9$,
3. finally, we obtain the AES key $K$ by exploiting the faulty ciphertexts obtained by introducing a fault on $M^8$ before entering Round 9 and by using the 8 bytes of $K^9$ disclosed in steps 1 and 2.

In smartcard implementations, each round key is computed on-the-fly. In the following section, "attack on $K^i$" means that the correct $i^{th}$ round key has been used for the cipher and that a fault has been induced on this round key before computing the $i + 1^{th}$ round key which is a faulty round key.

## 4.1 DFA Attack on $K^9$

We suppose that we know both the correct ciphertext $C$ and a faulty ciphertext $D$ of the same plaintext $M$ and that the fault occurs on one of the bytes of $K^9$ just before computing $K^{10}$ as shown in figure 3, where the shaded squares represent the bytes affected by the fault.

We want the fault to occur on one of the last 4 bytes of $K^9$. In that case, two of the last 4 bytes of the faulty ciphertext will be different from those of the correct ciphertext. We must hence check if this condition is true: if it is not, we abandon this faulty ciphertext and we generate another faulty ciphertext with a fault on $K^9$ and we test it again.

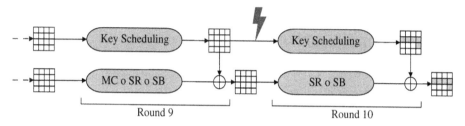

**Fig. 3.** Fault on the $14^{\text{th}}$ byte of the penultimate round key $K^9$

Now, we will see that it is possible to identify:

- the position $j$ of the byte on which the fault occurred
- and the value $e_j$ of this fault.

If we suppose that a fault $e_j$ occurs on the $j^{\text{th}}$ byte of $K^9$ ($12 \leq j \leq 15$) just before the Final Round, there will only be one non-zero byte in the first 4 bytes of $C \oplus D$. If we denote this byte the $k^{\text{th}}$ ($0 \leq k \leq 3$), $j$ is then defined by

$$j = (k + 1 \bmod 4) + 12 \tag{7}$$

By computing $C \oplus D$, we determine $k$ and thus obtain $j$.

By definition, we have:

$$\forall i \in \{0, ..., 15\}, \quad C_i = SubByte(M^9_{ShiftRow^{-1}(i)}) \oplus K^{10}_i \tag{8}$$

More precisely:

- if $i = 0$:

$$C_i = SubByte(M^9_{ShiftRow^{-1}(i)}) \oplus SubByte(K^9_{(i+1 \bmod 4)+12}) \oplus K^9_i \oplus 0x36 \tag{9}$$

- if $i \in \{1, 2, 3\}$:

$$C_i = SubByte(M^9_{ShiftRow^{-1}(i)}) \oplus SubByte(K^9_{(i+1 \bmod 4)+12}) \oplus K^9_i \tag{10}$$

We also have for the faulty ciphertext:

$$D_j = SubByte(M^9_{ShiftRow^{-1}(j)}) \oplus K^{10}_j \oplus e_j \tag{11}$$

and

- if $k = 0$:

$$D_k = SubByte(M^9_{ShiftRow^{-1}(k)}) \oplus SubByte(K^9_j \oplus e_j) \oplus K^9_k \oplus 0x36 \qquad (12)$$

- if $k \in \{1, 2, 3\}$:

$$D_k = SubByte(M^9_{ShiftRow^{-1}(k)}) \oplus SubByte(K^9_j \oplus e_j) \oplus K^9_k \qquad (13)$$

It is easy to see, from (8) and (11), that the value of the fault $e_j$ is equal to $C_j \oplus D_j$.

We have now identify the position $j$ of the byte on which the fault occurred and the value $e_j$ of this fault. Let us see how to use this information to obtain the value of $K^9_j$.

From (9), (10), (12) and (13), we have the equation

$$C_k \oplus D_k = SubByte(K^9_j) \oplus SubByte(K^9_j \oplus e_j) \qquad (14)$$

We know the value of $C_k \oplus D_k$ and the value of $e_j$. So, we search the possible values $x \in \{0, ..., 255\}$ which satisfy the equation

$$C_k \oplus D_k = SubByte(x) \oplus SubByte(x \oplus e_j) \qquad (15)$$

We obtain $K^9_j$ and $K^9_j \oplus e_j$ as solutions to (15). So, if we obtain another faulty ciphertext with a fault $e'_j$ $(e'_j \neq e_j)$ which occurs on the same byte $j$ of $K^9$, we obtain $K^9_j$ and $K^9_j \oplus e'_j$ as solutions. This allows us to deduce the value of $K^9_j$ because it is the only value that appears in both solution.

With this attack, we obtain the values of the last 4 bytes ($K^9_{12}$ to $K^9_{15}$) of the round key $K^9$ with 32 faulty ciphertexts on average.

## 4.2    Attack on $K^8$

Now, we will see how to obtain the 4 bytes $K^9_8$ to $K^9_{11}$. We use faulty ciphertexts obtained when the fault $e_j$ occurred on one byte of $K^8$ (lets say the $j^{th}$ byte) before Round 9.

We want the fault to occur on one of the last 4 bytes of $K^8$. If it is the case, there will only be one zero byte in the last 4 bytes of $C \oplus D$. So we test this

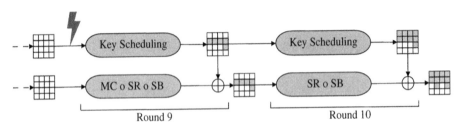

**Fig. 4.** Fault on the $14^{th}$ byte of the antepenultimate round key $K^8$

condition and if it is false, we generate another faulty ciphertext with a fault induced on $K^8$ and we test it again.

As in section 4.1, we will:

- identify the position $j$ of the byte on which the fault occurred
- and obtain the value $e_j$ of this fault.

If we denote by $l$ the position of the zero byte in the last 4 bytes of $C \oplus D$ ($12 \le l \le 15$), $j$ is then defined by

$$j = (l - 1 \bmod 4) + 12 \tag{16}$$

Now, we know on which byte of $K^8$ the fault occurred.

We have, for the faulty ciphertext $D$:

$$\begin{cases} D_j = SubByte(M^9_{ShiftRow^{-1}(j)}) \oplus K^{10}_j \oplus e_j & \text{if } j \ne 12 \\ D_j = SubByte(M^9_{ShiftRow^{-1}(j)} \oplus e_j) \oplus K^{10}_j \oplus e_j & \text{if } j = 12 \end{cases} \tag{17}$$

and for the correct ciphertext:

$$C_i = SubByte(M^9_{ShiftRow^{-1}(i)}) \oplus K^{10}_i \quad \forall i \in \{0, ..., 15\} \tag{18}$$

- If $j \ne 12$, we easily obtain the value of $e_j$ which is equal to $C_j \oplus D_j$.
- But, if $j = 12$, the *ShiftRows* transformation does not affect the $12^{\text{th}}$ byte and we cannot directly obtain the value of the fault $e_j$. We only know that

$$C_j \oplus D_j = SubByte(a) \oplus SubByte(a \oplus e_j) \oplus e_j \tag{19}$$

for a certain 8-bit value $a$. In this case, we guess the fault $e_j$ and we look for a value $a$ which satisfies (19). If such a value exists, we assume that our guess may be correct and we keep it as a possible value for the fault $e_j$. We obtain between 107 and 146 different possible values for $e_j$ depending on the value of $C_j \oplus D_j$; the average is about 127.

Now, we have identify the position $j$ of the byte on which the fault occurred and the value $e_j$ of this fault if $j \ne 12$ or a set of possible values if $j = 12$. Let us see how to use this information to obtain the value of $K^8_j$.

If we induce a fault on $K^8_j$ ($12 \le j \le 15$), the 4 bytes of the faulty $9^{\text{th}}$ round key at position $(j - 1 \bmod 12) + 4n$, $n \in \{0, 1, 2, 3\}$, are different from the bytes at the same position of the correct $9^{\text{th}}$ round key $K^9$. These four differences between the correct and the faulty $9^{\text{th}}$ round key are equal and we denote this difference $f_j$.

If we denote $k = (j - 1 \bmod 4) + 12$, we have $K^9_k \oplus f_j$ as the value of the $k^{\text{th}}$ byte of the faulty $9^{\text{th}}$ round key.

So, we have:
- if $j = 14$:

$$D_{j-2 \bmod 4} = SubByte(M^9_{ShiftRow^{-1}(j-2 \bmod 4)}) \oplus SubByte(K^9_k \oplus f_j) \\ \oplus K^9_{j-2 \bmod 4} \oplus 0x36 \tag{20}$$

- if $j \in \{12, 13, 15\}$:

$$D_{j-2 \bmod 4} = SubByte(M^9_{ShiftRow^{-1}(j-2 \bmod 4)}) \oplus SubByte(K^9_k \oplus f_j) \qquad (21)$$
$$\oplus K^9_{j-2 \bmod 4}$$

And we obtain from (9), (10), (20) and (21):

$$C_{j-2 \bmod 4} \oplus D_{j-2 \bmod 4} = SubByte(K^9_k) \oplus SubByte(K^9_k \oplus f_j) \qquad (22)$$

As we know the value of $K^9_k$ from the previous attack (section 4.1), we can easily find the value of $f_j$ which satisfies (22).

Moreover, $K^9_{j-1 \bmod 4} \oplus f_j$ is the value of the $(j - 1 \bmod 12)^{\text{th}}$ byte of the faulty $9^{\text{th}}$ round key. So, we have for the faulty Key Scheduling:

- if $j = 13$:

$$SubByte(K^8_j \oplus e_j) \oplus K^8_{j-1 \bmod 4} \oplus 0x36 = K^9_{j-1 \bmod 4} \oplus f_j \qquad (23)$$

- if $j \in \{12, 14, 15\}$:

$$SubByte(K^8_j \oplus e_j) \oplus K^8_{j-1 \bmod 4} = K^9_{j-1 \bmod 4} \oplus f_j \qquad (24)$$

and for the correct Key Scheduling:

- if $j = 13$:

$$SubByte(K^8_j) \oplus K^8_{j-1 \bmod 4} \oplus 0x36 = K^9_{j-1 \bmod 4} \qquad (25)$$

- if $j \in \{12, 14, 15\}$:

$$SubByte(K^8_j) \oplus K^8_{j-1 \bmod 4} = K^9_{j-1 \bmod 4} \qquad (26)$$

We obtain from (23), (24), (25) and (26):

$$f_j = SubByte(K^8_j \oplus e_j) \oplus SubByte(K^8_j) \qquad (27)$$

With the value of $f_j$ previously obtained from (22), we find all the possible values $K^8_j$ which satisfy (27).

As in section 3, we use a counting method in order to find the correct $K^8_j$. The right $K^8_j$ can be obtained quickly when $j \neq 12$ because we know the value of the fault $e_j$. However, if $j = 12$ it is more difficult because there are many possible values for $e_j$ (between 107 and 146). Although we need more faulty ciphertexts to determine $K^8_{12}$ than to determine $K^8_{13}$, $K^8_{14}$ or $K^8_{15}$, the number required is relatively low. We need approximately 13 faulty ciphertexts from the same plaintext to obtain $K^8_{12}$ and only 2 to obtain $K^8_{13}$, $K^8_{14}$ or $K^8_{15}$ (by using simulation, we find that we have a 90% chance of success to determine $K^8_{12}$ if we use 10 faulty ciphertexts and this percentage grows up to 99% if we use 13 faulty ciphertexts).

Finally, to obtain $K^9_8$, $K^9_9$, $K^9_{10}$ and $K^9_{11}$, we use the following formula:

$$K^9_i = K^8_{i+4} \oplus K^9_{i+4} \qquad \forall i \in \{8, ..., 11\} \qquad (28)$$

At this step, we have obtained the last 8 bytes of the penultimate round key $K^9$ by using about 240 faulty ciphertexts.

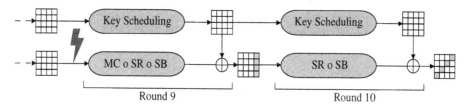

**Fig. 5.** Fault on the $11^{\text{th}}$ byte of $M^8$

### 4.3   DFA Attack on $M^8$

Before entering Round 9, we assume that a fault on one byte of $M^8$ has been induced. As we have determined the last 8 bytes of $K^9$, we want the fault to occur on a byte of $M^8$ which will be XORed with one of the last 8 bytes of $K^9$ after $MC \circ SR \circ SB$. Due to the *ShiftRows* and *MixColumns* transformations, we know that if we induce a fault on $M_{12}^8$, $M_1^8$, $M_6^8$ or on $M_{11}^8$ (resp. on $M_8^8$, $M_{13}^8$, $M_2^8$ or on $M_7^8$), the result of these bytes after $MC \circ SR \circ SB$ will be XORed with $K_{12}^9$ to $K_{15}^9$ (resp. $K_8^9$ to $K_{11}^9$). So, we want a fault to occur on one of these 8 bytes of $M^8$ and to test if this happens, we look at the faulty ciphertext: if only the 4 bytes $(D_{12}, D_9, D_6, D_3)$ (resp. $(D_8, D_5, D_2, D_{15})$) differ from $(C_{12}, C_9, C_6, C_3)$ (resp. $(C_8, C_5, C_2, C_{15})$) of the correct ciphertext, this shows that the fault occurred on one of the 4 bytes $(M_{12}^8, M_1^8, M_6^8, M_{11}^8)$ (resp. $(M_8^8, M_{13}^8, M_2^8, M_7^8)$).

In the following, let $(D_{12}, D_9, D_6, D_3)$ be different from $(C_{12}, C_9, C_6, C_3)$. We guess the fault $e_j$ $(1 \le e_j \le 255)$ and we list all the 4-byte values $V$ which verify one of the following equations:

$$
\begin{aligned}
SB(MC(V) \oplus K_{12-15}^9) \oplus SB(MC(V \oplus (0,0,0,e_j)) \oplus K_{12-15}^9) = TR_{12-15} \\
SB(MC(V) \oplus K_{12-15}^9) \oplus SB(MC(V \oplus (0,0,e_j,0)) \oplus K_{12-15}^9) = TR_{12-15} \\
SB(MC(V) \oplus K_{12-15}^9) \oplus SB(MC(V \oplus (0,e_j,0,0)) \oplus K_{12-15}^9) = TR_{12-15} \\
SB(MC(V) \oplus K_{12-15}^9) \oplus SB(MC(V \oplus (e_j,0,0,0)) \oplus K_{12-15}^9) = TR_{12-15}
\end{aligned}
\tag{29}
$$

where $K_{12-15}^9$ denotes the 4-byte value $(K_{12}^9, K_{13}^9, K_{14}^9, K_{15}^9)$ and $TR_{12-15}$ the 4-byte value $(C \oplus D)_{ShiftRow(12-15)} = (C_{12} \oplus D_{12}, C_9 \oplus D_9, C_6 \oplus D_6, C_3 \oplus D_3)$.

So, if we apply the same reasoning to another faulty ciphertext which differs from the correct ciphertext on $(D_{12}, D_9, D_6, D_3)$, we obtain another list of 4-byte values. There will only be one 4-byte value present in both lists and this will be the correct value of the last 4 bytes of the temporary result before the *MixColumns* transformation in Round 9.

Proceeding in the same way with two different faulty ciphertexts in which $(D_8, D_5, D_2, D_{15})$ differ from $(C_8, C_5, C_2, C_{15})$, we obtain the correct $8^{\text{th}}$ to $11^{\text{th}}$ bytes of the temporary result before the *MixColumns* transformation in Round 9.

Having now identified the last 8 bytes of the temporary cipher result before the *MixColumns* transformation in Round 9, we apply *MixColumns* to these 8 bytes. We then XOR the result with the corresponding bytes of $K^9$ (i.e. $K_8^9$ to $K_{15}^9$) and we apply $SR \circ SB$. This result is a part of the correct temporary result before the XOR with $K^{10}$. So, we XOR it with the corresponding bytes of the ciphertext $C$ to obtain the bytes $K_2^{10}$, $K_3^{10}$, $K_5^{10}$, $K_6^{10}$, $K_8^{10}$, $K_9^{10}$, $K_{12}^{10}$ and $K_{15}^{10}$. Using the known bytes of $K^9$, we obtain 6 other bytes of $K^{10}$ by the following relations:

$$
\begin{aligned}
K_{13}^{10} &= K_9^{10} \oplus K_{13}^9 \\
K_{11}^{10} &= K_{15}^{10} \oplus K_{15}^9 \\
K_{10}^{10} &= K_6^{10} \oplus K_{10}^9 \\
K_{14}^{10} &= K_{10}^{10} \oplus K_{14}^9 \\
K_7^{10} &= K_{11}^{10} \oplus K_{11}^9 \\
K_4^{10} &= K_8^{10} \oplus K_8^9
\end{aligned}
\tag{30}
$$

Finally, we find the last 2 unknown bytes of $K^{10}$ by a very fast exhaustive search and we obtain the AES key from $K^{10}$ by applying the inverse of the Key Scheduling.

Theoretically, we obtain the full AES key by using less than 250 faulty ciphertexts.

## 5   Remark

The previous number of required faulty ciphertexts was determined by supposing that the fault location cannot be chosen, i.e. the position of the fault is uniformly distributed among the 16 bytes of a chosen temporary result. If we suppose that we can choose the byte where the fault is induced, we need on average 35 faulty ciphertexts to recover the secret key by using our bit-fault attack and only 31 faulty ciphertexts by using our byte-fault attack (we need 8 faulty ciphertexts to perform the fault attack described in section 4.1, 19 to perform the one described in section 4.2 and 4 to perform the one described in section 4.3).

## 6   In Practice

We implemented the algorithmic part of the second attack on an AES-128 and, by simulating faults on random bytes of $K^8$, $K^9$ and $M^9$, we found the whole AES key by using 250 faulty ciphertexts. This was easily done on a computer but we were yet to discover if our second attack could be successfully put into practice on a smart card.

By using a microscope, a modified camera flash and a computer, we attacked an AES-128 on an 8-bit smart card (to make the attack easier, we used a known AES code). Firstly, we had to find out where the light flash was most efficient on the surface of the chip and then we had to synchronize the flash with the operations we wanted to disturb.

We even succeeded in inducing a fault for nearly every execution of the AES, we needed a lot of tries to obtain a "good" faulty ciphertext. Indeed, most of the time, the induced fault affected 4 or 8 bytes of the temporary result.

To recover the key, we needed numerous tries: more than 1000 AES executions were required.

If we had had a laser we could have shortened the length of the flash and hence obtained a "good" faulty ciphertext more frequently by disturbing the chip for a very short time, i.e. during the treatment of only one byte.

This experience demonstrates that AES on smart cards must now be implemented not only with SPA/DPA countermeasures but also with DFA countermeasures.

## 7    Conclusion

Although DFA on the DES is a well-known attack, it is impossible to directly apply Biham and Shamir's attack to the AES as the latter does not have the Feistel Structure. This paper extends the operative field of differential fault attacks by describing how to perform two different DFA attacks on the AES. Each of these attacks allow us to obtain the full AES key in the case of a 128-bit key length. We note that it is possible to put the second attack into practice on smart cards. However, it is easy to avoid both attacks. For example, this can be done by doubling the last two rounds and by checking if the two outputs are equal.

## Acknowledgments

I would like to thank Mathieu Ciet for his valuable comments as well as Erik Knudsen for many helpful discussions. The practical attack would never have been possible without the help of Hugues Thiebeauld. Finally, I am really grateful to Julia Bradley for her help and support during the writing of this paper.

## References

1. R. Anderson and M. Kuhn. Tamper Resistance - a Cautionary Note. In *Proceedings of the* $2^{nd}$ *USENIX Workshop on Electronic Commerce*, pages 1–11, 1996.
2. R. Anderson and M. Kuhn. Low cost attacks on tamper resistant devices. In B. Christianson, B. Crispo, T. Mark, A. Lomas, and M. Roe, editors, $5^{th}$ *Security Protocols Workshop*, volume 1361 of *Lecture Notes in Computer Science*, pages 125–136. Springer-Verlag, 1997.
3. I. Biehl, B. Meyer, and V. Müller. Differential Fault Analysis on Elliptic Curve Cryptosystems. In M. Bellare, editor, *Advances in Cryptology – CRYPTO 2000*, volume 1880 of *Lecture Notes in Computer Science*, pages 131–146. Springer-Verlag, 2000.
4. E. Biham and A. Shamir. Differential Fault Analysis of Secret Key Cryptosystem. In B.S. Kalisky Jr., editor, *Advances in Cryptology – CRYPTO '97*, volume 1294 of *Lecture Notes in Computer Science*, pages 513–525. Springer-Verlag, 1997.
5. J. Blömer and J.-P. Seifert. Fault based cryptanalysis of the Advanced Encryption Standard. In R.N. Wright, editor, *Financial Cryptography – FC 2003*, volume 2742 of *Lecture Notes in Computer Science*. Springer-Verlag, 2003.

6.  D. Boneh, R.A. DeMillo, and R.J. Lipton. On the Importance of Checking Cryptographic Protocols for Faults. In W. Fumy, editor, *Advances in Cryptology – EUROCRYPT '97*, volume 1233 of *Lecture Notes in Computer Science*, pages 37–51. Springer-Verlag, 1997.
7.  M. Ciet and M. Joye. Elliptic Curve Cryptosystems in the Presence of Permanent and Transient Faults. In *Designs, Codes and Cryptography*, 2004. To appear.
8.  J. Daemen and V. Rijmen. *The Design of Rijndael*. Springer-Verlag, 2002.
9.  A.K. Lenstra. Memo on RSA Signature Generation in the Presence of Faults. Manuscript, 1996. Available from the author at akl@Lucent.com.
10. D.P. Maher. Fault Induction Attacks, Tamper Resistance, and Hostile Reverse Engineering in Perspective. In R. Hirschfeld, editor, *Financial Cryptography – FC '97*, volume 1318 of *Lecture Notes in Computer Science*, pages 109–121. Springer-Verlag, 1997.
11. National Institute of Standards and Technology. *FIPS PUB 197: Advanced Encryption Standard*, 2001.
12. G. Piret and J.-J. Quisquater. A Differential Fault Attack Technique Against SPN Structures, with Application to the AES and KHAZAD. In C.D. Walter, Ç.K. Koç, and C. Paar, editors, *Cryptographic Hardware and Embedded Systems – CHES 2003*, volume 2779 of *Lecture Notes in Computer Science*, pages 77–88. Springer-Verlag, 2003.
13. S. Skorobogatov and R. Anderson. Optical Fault Induction Attack. In B. Kaliski Jr., Ç.K. Koç, and C. Paar, editors, *Cryptographic Hardware and Embedded Systems – CHES 2002*, volume 2523 of *Lecture Notes in Computer Science*, pages 2–12. Springer-Verlag, 2002.

## A     The First Attack in More Details

If a message $M$ is ciphered by using an AES-128 and if a one-bit fault $e_j$ is induced on $M_j^9$, we obtain a faulty ciphertext $D$. We then have the following equation:

$$C_{ShiftRow(j)} \oplus D_{ShiftRow(j)} = SubByte(M_j^9) \oplus SubByte(M_j^9 \oplus e_j) \qquad (31)$$

For each faulty ciphertext we perform $8.2^8$ tests, i.e. for all values of $x$ between 0 and 255 and for $e_j \in \{0x01, 0x02, 0x04, 0x08, 0x10, 0x20, 0x40, 0x80\}$, we test if the following equality holds :

$$C_{ShiftRow(j)} \oplus D_{ShiftRow(j)} = SubByte(x) \oplus SubByte(x \oplus e_j) \qquad (32)$$

There is no solution to (32) if $C_{ShiftRow(j)} \oplus D_{ShiftRow(j)} = 185$, so this value can be excluded right away. By consecutively fixing the left hand side of (32) with the 254 possible values $\{1, .., 255\} \backslash \{185\}$ and by testing all possible pairs $(x, e_j)$, we find that the number of possible values for $M_j^9$ varies from 2 to 14; the average is about 8.

If we assume that we are in the worst case, then we obtain 14 possible values for $M_j^9$ for each faulty ciphertext.

If we obtain another faulty ciphertext with an induced fault on $M_j^9$ we obtain another set of possible values for $M_j^9$. In each set we have the correct value of $M_j^9$, so to identify this value the other 13 values must be different from each other.

If we denote by $A$ the set of these 13 values obtained with the first faulty ciphertext and by $B$ the set of the possible values obtained with the second faulty ciphertext except the correct value of $M_j^9$, we have only one possible value left for $M_j^9$ with probability :

$$
\begin{aligned}
P_2 &= P(A \cap B = \emptyset) \\
&= P(|A \cap B| = 0) \\
&= \frac{\dbinom{255}{13} * \dbinom{255 - 13}{13}}{\dbinom{255}{13}^2} \\
&\simeq 50\%
\end{aligned}
\tag{33}
$$

With a third faulty ciphertext with an induced fault on $M_j^9$ we obtain yet another set of 14 possible values for $M_j^9$. If we denote by $C$ this set without the correct value of $M_j^9$, we have only one possible value left for $M_j^9$ with probability :

$$
\begin{aligned}
P_3 &= P(A \cap B \cap C = \emptyset) \\
&= P(|A \cap B \cap C| = 0) \\
&= \sum_{k=0}^{min\{|A|,|B|\}} P(|A \cap B| = k, \ |A \cap B \cap C| = 0) \\
&= \sum_{k=0}^{13} P(|A \cap B| = k) * P(|A \cap B \cap C| = 0 \ / \ |A \cap B| = k) \\
&= \sum_{k=0}^{13} \frac{\dbinom{255}{13} * \dbinom{13}{k} * \dbinom{255 - 13}{13 - k}}{\dbinom{255}{13}^2} * \frac{\dbinom{255}{k} * \dbinom{255 - k}{13}}{\dbinom{255}{k} * \dbinom{255}{13}} \\
&\simeq 97\%
\end{aligned}
\tag{34}
$$

# Refined Analysis of Bounds Related to Linear and Differential Cryptanalysis for the AES

Liam Keliher

Department of Mathematics and Computer Science,
Mount Allison University,[**]
Sackville, New Brunswick, Canada
lkeliher@mta.ca

**Abstract.** The best upper bounds on the maximum expected linear probability (MELP) and the maximum expected differential probability (MEDP) for the AES, due to Park et al. [23], are $1.075 \times 2^{-106}$ and $1.144 \times 2^{-111}$, respectively, for $T \geq 4$ rounds. These values are simply the $4^{\text{th}}$ powers of the best upper bounds on the MELP and MEDP for $T = 2$ [3, 23]. In our analysis we first derive nontrivial *lower* bounds on the 2-round MELP and MEDP, thereby trapping each value in a small interval; this demonstrates that the best 2-round upper bounds are quite good. We then prove that these same 2-round upper bounds are not tight—and therefore neither are the corresponding upper bounds for $T \geq 4$. Finally, we show how a modified version of the KMT2 algorithm (or its dual, KMT2-DC), due to Keliher et al. (see [8]), can potentially improve any existing upper bound on the MELP (or MEDP) for any SPN. We use the modified version of KMT2 to improve the upper bound on the AES MELP to $1.778 \times 2^{-107}$, for $T \geq 8$.

**Keywords:** AES, Rijndael, SPN, provable security, linear cryptanalysis, differential cryptanalysis, MELP, MEDP, KMT2, KMT2-DC.

## 1 Introduction

During the past few years, several papers have appeared dealing with the provable security of substitution-permutation network (SPN) block ciphers against linear and differential cryptanalysis [3, 6, 7, 9, 10, 11, 12, 22, 23, 24]. Most of these results have been applied to the Advanced Encryption Standard (AES) [5]—each new result has demonstrated greater provable security against one or both of these attacks.

Exhibiting provable security against linear and differential cryptanalysis requires proving that the maximum expected linear probability (MELP) and the maximum expected differential probability (MEDP), respectively, are small over

[**] This work was funded by the Natural Sciences and Engineering Research Council of Canada (NSERC).

H. Dobbertin, V. Rijmen, A. Sowa (Eds.): AES 2004, LNCS 3373, pp. 42–57, 2005.

**Table 1.** Previous upper bounds on the MELP and MEDP for the AES

| MELP | MEDP | Range of rounds |
|------|------|-----------------|
| $2^{-24}$ [7] | $2^{-24}$ [7] | $T \geq 2$ |
| $2^{-75}$ [9] | $2^{-75}$ [10] | $T \geq 7$ |
| $2^{-92.4}$ [11, 12] | $2^{-95.1}$ [12] | $T \geq 9$ |
| $2^{-96}$ [24] | $2^{-96}$ [24] | $T \geq 4$ |
| $1.06 \times 2^{-96}$ [22] | $1.06 \times 2^{-96}$ [22] | $T \geq 4$ |
| $1.075 \times 2^{-106}$ [23] | $1.144 \times 2^{-111}$ [23] | $T \geq 4$ |

$T$ core cipher rounds (typically $T = R - 1$ or $T = R - 2$, where $R$ is the total number of rounds). Since exact computation of these values often appears
to be infeasible, researchers have focused on bounds. A sufficiently small upper bound corresponds to a data complexity that is prohibitively large, since
the data complexity is proportional to the inverse of the corresponding MELP /
MEDP [19, 20]. Note that bounds often appear in pairs—one each for the MELP
and MEDP—because of the well-known duality between linear and differential
cryptanalysis [1, 17]. Table 1 summarizes the upper bounds that have been derived for the AES prior to the current paper. [1] [2]

The best upper bounds in Table 1 (last row), due to Park et al., are in fact
the $4^{\text{th}}$ powers of the best upper bounds on the MELP and MEDP for $T = 2$,
namely $\frac{48,193,441}{2^{52}} \approx 1.44 \times 2^{-27}$ and $\frac{79}{2^{34}} \approx 1.23 \times 2^{-28}$, respectively [3, 23]. In
fact, Park et al. show that the $4^{\text{th}}$ power of *any* upper bound on the 2-round
MELP / MEDP for the AES is an upper bound on the MELP / MEDP for
$T \geq 4$ (this also follows from the work of Sano et al. [24]). Therefore the 2-round
MELP and MEDP are important values for analyzing the resistance of the AES
to linear and differential cryptanalysis.

In this paper we first derive nontrivial *lower* bounds on the 2-round MELP
and MEDP for the AES, namely $1.638 \times 2^{-28}$ and $1.656 \times 2^{-29}$, respectively,
thereby trapping each value in a small interval; this demonstrates that the best
2-round upper bounds are quite good.[3] Second, we prove that these same 2-round
upper bounds are not tight—and therefore neither are the corresponding upper
bounds for $T \geq 4$. Third, we show how a modified version of the KMT2 algorithm
(or its dual, KMT2-DC), due to Keliher et al. (see [8]), can potentially improve
any existing upper bound on the MELP (or MEDP) for any SPN. We use the
modified version of KMT2 to improve the upper bound on the AES MELP

---

[1] The results in [7] were not applied to the AES, but the values in the first row of
Table 1 are the upper bounds that would have resulted.

[2] The almost identical bounds in [24] and [22] were apparently obtained independently.

[3] After the presentation of this paper, we learned that the same lower bounds had
previously been obtained by Chun et al. [3].

to $1.778 \times 2^{-107}$, for $T \geq 8$. (The KMT2 / KMT2-DC algorithm computes upper bounds on the MELP / MEDP "from scratch"; the modification involves incorporating existing upper bounds that are superior to those computed directly by KMT2 / KMT2-DC in order to refine the former.)

## 1.1   The Advanced Encryption Standard (AES)

The *Advanced Encryption Standard* (AES) is a U.S. block cipher standard selected in 2000 after an open submission and evaluation process. The AES is the SPN *Rijndael*, designed by Joan Daemen and Vincent Rijmen [5]. A single AES round (minus the subkey mixing) is depicted in Figure 1.

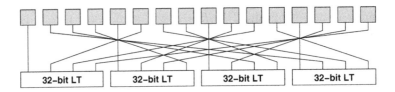

**Fig. 1.** One AES round

The AES has a block size of 128 bits. The substitution stage consists of 16 identical $8 \times 8$ s-boxes (the same s-box is used in all rounds). The linear transformation comprises two steps: a byte permutation, and the parallel application of four copies of a maximally diffusive 32-bit linear transformation (see Remark 4). The number of rounds varies according to the key length as follows: 128-bit key $\Rightarrow$ 10 rounds, 192-bit key $\Rightarrow$ 12 rounds, 256-bit key $\Rightarrow$ 14 rounds.

## 1.2   Assumption of Independent Subkeys

In analyzing the resistance of block ciphers to linear and differential cryptanalysis, it is standard to assume that each subkey is chosen independently and uniformly from the set of all possible subkeys.[4] We adopt this approach. Because of the complexities introduced by most key schedules, the values relevant to linear and differential cryptanalysis are rarely calculated for the true distribution of subkeys—this remains an interesting and largely unexplored area of study.

## 2   Linear and Differential Cryptanalysis

Linear and differential cryptanalysis are generally considered to be the most powerful attacks on block ciphers. Linear cryptanalysis, due to Matsui [16], is a known-plaintext attack that exploits the existence of relatively large *expected*

---

[4] Some authors use AES* to denote the AES modified by this assumption.

*linear probability* (ELP) values over $T$ core cipher rounds. Differential cryptanalysis, due to Biham and Shamir [2], is a chosen-plaintext attack that exploits the existence of relatively large *expected differential probability* (EDP) values over $T$ core rounds. Typical values of interest are $T = R - 1$ and $T = R - 2$.

The remainder of this section deals with background concepts related to linear and differential cryptanalysis of SPNs. We use $N$ to denote the block size, $n$ to denote the s-box input/output size, and $M$ to denote the number of s-boxes per round (so $M = \frac{N}{n}$). We assume that the same linear transformation and sequence of s-boxes are used in each round (the s-boxes within a round may or many not be identical). It is easy to generalize to the situation in which the linear transformation and s-boxes differ from round to round.

## 2.1   Linear and Differential Probability

**Definition 1.** *Let* $B : \{0,1\}^d \rightarrow \{0,1\}^d$, *and let* $\mathbf{a}, \mathbf{b}, \Delta\mathbf{x}, \Delta\mathbf{y} \in \{0,1\}^d$ *be fixed. If* $\mathbf{X} \in \{0,1\}^d$ *is a uniformly distributed random variable, then the* linear probability $LP(\mathbf{a}, \mathbf{b})$ *and the* differential probability $DP(\Delta\mathbf{x}, \Delta\mathbf{y})$ *are defined as*

$$LP(\mathbf{a}, \mathbf{b}) = (2 \cdot \text{Prob}_{\mathbf{X}} \{\mathbf{a} \bullet \mathbf{X} = \mathbf{b} \bullet B(\mathbf{X})\} - 1)^2$$
$$DP(\Delta\mathbf{x}, \Delta\mathbf{y}) = \text{Prob}_{\mathbf{X}} \{B(\mathbf{X}) \oplus B(\mathbf{X} \oplus \Delta\mathbf{x}) = \Delta\mathbf{y}\}.$$

*If $B$ is parameterized by a key,* $\mathbf{k}$, *we write* $LP(\mathbf{a}, \mathbf{b}; \mathbf{k})$ *and* $DP(\Delta\mathbf{x}, \Delta\mathbf{y}; \mathbf{k})$, *respectively, and the* expected linear probability $ELP(\mathbf{a}, \mathbf{b})$ *and* expected differential probability *are defined as*

$$ELP(\mathbf{a}, \mathbf{b}) = E_{\mathbf{K}} [LP(\mathbf{a}, \mathbf{b}; \mathbf{K})]$$
$$EDP(\Delta\mathbf{x}, \Delta\mathbf{y}) = E_{\mathbf{K}} [DP(\Delta\mathbf{x}, \Delta\mathbf{y}; \mathbf{K})],$$

*where* $\mathbf{K}$ *is a random variable uniformly distributed over the space of keys.*

We view LP, ELP, DP, or EDP values as entries in a $2^d \times 2^d$ table in the obvious way. The values $\mathbf{a}$ and $\mathbf{b}$ in Definition 1 are called input and output *masks,* and the values $\Delta\mathbf{x}$ and $\Delta\mathbf{y}$ are called input and output *differences.* For our purposes, the mapping $B$ in Definition 1 will be bijective, and will be an s-box, a single encryption round, or a sequence of consecutive encryption rounds.

**Lemma 1.** *Let* $B : \{0,1\}^d \rightarrow \{0,1\}^d$ *be bijective, and let* $\mathbf{a}, \mathbf{b}, \Delta\mathbf{x}, \Delta\mathbf{y} \in \{0,1\}^d$. *Then*

$$\sum_{\mathbf{u} \in \{0,1\}^d} LP(\mathbf{a}, \mathbf{u}) = \sum_{\mathbf{u} \in \{0,1\}^d} LP(\mathbf{u}, \mathbf{b}) = 1 \qquad (1)$$

$$\sum_{\Delta\mathbf{u} \in \{0,1\}^d} DP(\Delta\mathbf{x}, \Delta\mathbf{u}) = \sum_{\Delta\mathbf{u} \in \{0,1\}^d} DP(\Delta\mathbf{u}, \Delta\mathbf{y}) = 1. \qquad (2)$$

*Proof.* The proof of (1) derives directly from Parseval's Theorem [18]. The proof of (2) is trivial.

*Remark 1.* In what follows, terms such as "first round" and "last round" are relative to the $T$ rounds under consideration. Single-variable superscripts refer to individual rounds, e.g., $LP^t(\mathbf{a}, \mathbf{b}; \mathbf{k}^t)$ and $ELP^t(\mathbf{a}, \mathbf{b})$ are LP and ELP values, respectively, for round $t$ $(1 \leq t \leq T)$. Superscripts of the form $[i \ldots j]$ (with $i < j$) refer to a sequence of consecutive rounds viewed as a single unit, e.g., $EDP^{[1\ldots3]}(\Delta\mathbf{x}, \Delta\mathbf{y})$ is an EDP value over rounds $1\ldots3$.

## 2.2   Provable Security (MELP and MEDP)

Given $T \geq 2$ core rounds under consideration, the critical value for linear cryptanalysis is the *maximum expected linear probability* (MELP)[5]:

$$MELP = \max_{\mathbf{a},\mathbf{b}\in\{0,1\}^N\backslash\mathbf{0}} ELP^{[1\ldots T]}(\mathbf{a}, \mathbf{b}) . \tag{3}$$

The critical value for differential cryptanalysis is the *maximum expected differential probability* (MEDP):

$$MEDP = \max_{\Delta\mathbf{x},\Delta\mathbf{y}\in\{0,1\}^N\backslash\mathbf{0}} EDP^{[1\ldots T]}(\Delta\mathbf{x}, \Delta\mathbf{y}) . \tag{4}$$

For linear cryptanalysis / differential cryptanalysis, the data complexity of an attack with a given probability of success is proportional to the inverse of the MELP / MEDP. Therefore *provable security* can be claimed if this value is sufficiently small that the corresponding data complexity is prohibitive [19, 20].

## 2.3   Linear and Differential Characteristics

In general, for $T \geq 2$, it appears to be infeasible to compute the MELP or MEDP exactly for most SPNs. A traditional method of approximation involves the use of *characteristics*.

**Definition 2.** *A* linear characteristic / differential characteristic *for rounds* $1 \ldots T$ *is a* $(T+1)$-*tuple of* $N$-*bit masks / differences,* $\Omega = \langle \mathbf{a}^1, \mathbf{a}^2, \ldots, \mathbf{a}^T, \mathbf{a}^{T+1}\rangle$ / $\Omega = \langle \Delta\mathbf{x}^1, \Delta\mathbf{x}^2, \ldots, \Delta\mathbf{x}^T, \Delta\mathbf{x}^{T+1}\rangle$; *we view* $\mathbf{a}^t$ / $\Delta\mathbf{x}^t$ *and* $\mathbf{a}^{t+1}$ / $\Delta\mathbf{x}^{t+1}$ *as input and output masks / differences, respectively, for round* $t$ $(1 \leq t \leq T)$. *The corresponding* expected linear characteristic probability *(ELCP)* / expected differential characteristic probability *(EDCP) is defined as*

$$ELCP^{[1\ldots T]}(\Omega) = \prod_{t=1}^{T} ELP^t(\mathbf{a}^t, \mathbf{a}^{t+1}) \ /$$

$$EDCP^{[1\ldots T]}(\Omega) = \prod_{t=1}^{T} EDP^t(\Delta\mathbf{x}^t, \Delta\mathbf{x}^{t+1}) .$$

---

[5] A number of papers (including some by the author) use *maximum average linear hull probability* (MALHP), but MELP is more consistent with other related terminology.

*Remark 2.* For most SPNs, it is feasible to compute the values $ELP^t(\mathbf{a}^t, \mathbf{a}^{t+1})$ / $EDP^t(\Delta\mathbf{x}^t, \Delta\mathbf{x}^{t+1})$ (see Section 2.5), and therefore to compute $ELCP^{[1...T]}(\Omega)$ / $EDCP^{[1...T]}(\Omega)$.

**Using the Best Characteristic (Practical Security).** A *best* linear / differential characteristic is one that maximizes $ELCP^{[1...T]}(\Omega)$ / $EDCP^{[1...T]}(\Omega)$ (a best characteristic is not necessarily unique) . There are well-known (and relatively efficient) algorithms for finding best characteristics [17, 21]. Denote the best linear / differential characteristic by $\hat{\Omega}_L = \langle \hat{\mathbf{a}}^1, \hat{\mathbf{a}}^2, \ldots, \hat{\mathbf{a}}^T, \hat{\mathbf{a}}^{T+1} \rangle$ / $\hat{\Omega}_D = \langle \Delta\hat{\mathbf{x}}^1, \Delta\hat{\mathbf{x}}^2, \ldots, \Delta\hat{\mathbf{x}}^T, \Delta\hat{\mathbf{x}}^{T+1} \rangle$. The data complexity of linear / differential cryptanalysis is often estimated by assuming that

$$MELP = ELP^{[1...T]}(\hat{\mathbf{a}}^1, \hat{\mathbf{a}}^{T+1}) \approx ELCP^{[1...T]}(\hat{\Omega}_L) \ / \tag{5}$$

$$MEDP = EDP^{[1...T]}(\Delta\hat{\mathbf{x}}^1, \Delta\hat{\mathbf{x}}^{T+1}) \approx EDCP^{[1...T]}(\hat{\Omega}_D). \tag{6}$$

If the resulting data complexity is prohibitive, the cipher is *practically secure* [13].

## 2.4   Linear Hulls and Differentials

The concept of *linear hulls* is due to Nyberg [19]. The counterpart for differential cryptanalysis is the concept of *differentials*, due to Lai et al. [14].

**Definition 3.** *If $T \geq 2$ and $\mathbf{a}, \mathbf{b} \in \{0,1\}^N$ / $\Delta\mathbf{x}, \Delta\mathbf{y} \in \{0,1\}^N$, then the corresponding linear hull / differential, denoted $ALH(\mathbf{a}, \mathbf{b})$[6] / $DIFF(\Delta\mathbf{x}, \Delta\mathbf{y})$, is the set of all linear / differential characteristics for rounds $1 \ldots T$ having $\mathbf{a}$ / $\Delta\mathbf{x}$ as the first mask / difference and $\mathbf{b}$ / $\Delta\mathbf{y}$ as the last mask / difference, i.e., all linear / differential characteristics of the form*

$$\Omega = \langle \mathbf{a}, \mathbf{a}^2, \mathbf{a}^3, \ldots, \mathbf{a}^T, \mathbf{b} \rangle \ / \ \ \Omega = \langle \Delta\mathbf{x}, \Delta\mathbf{x}^2, \Delta\mathbf{x}^3, \ldots, \Delta\mathbf{x}^T, \Delta\mathbf{y} \rangle.$$

**Theorem 1 ([19, 14]).** *Let $\mathbf{a}, \mathbf{b} \in \{0,1\}^N$. Then*

$$ELP^{[1...T]}(\mathbf{a}, \mathbf{b}) = \sum_{\Omega \in ALH(\mathbf{a}, \mathbf{b})} ELCP^{[1...T]}(\Omega)$$

$$EDP^{[1...T]}(\Delta\mathbf{x}, \Delta\mathbf{y}) = \sum_{\Omega \in DIFF(\Delta\mathbf{x}, \Delta\mathbf{y})} EDCP^{[1...T]}(\Omega).$$

It follows from Theorem 1 that the approximation in (5) / (6) does not hold in general, since $ELP^{[1...T]}(\mathbf{a}, \mathbf{b})$ / $EDP^{[1...T]}(\Delta\mathbf{x}, \Delta\mathbf{y})$ is seen to be the sum of (a large number of) terms $ELCP^{[1...T]}(\Omega)$ / $EDCP^{[1...T]}(\Omega)$, and therefore, in general, the ELCP / EDCP of any characteristic will be strictly *less than* the corresponding ELP / EDP value. Further, the MELP / MEDP may not be equal to (i.e., may be strictly greater than) the ELP / EDP associated with any best characteristic. This situation may result in an overestimation of the data complexity—beneficial for an attacker, but problematic for a cipher designer.

---

[6] Nyberg originally used *approximate linear hull,* hence the abbreviation ALH.

## 2.5    Active S-Boxes and Branch Numbers

Let $\mathcal{L}$ denote the SPN linear transformation represented as an invertible $N \times N$ binary matrix, i.e., if $\mathbf{x}, \mathbf{y} \in \{0,1\}^N$ are the input and output, respectively, for the linear transformation, then $\mathbf{y} = \mathcal{L}\mathbf{x}$ (view $\mathbf{x}$ and $\mathbf{y}$ as column vectors).

**Lemma 2 ([5]).** *If* $\mathbf{a} \in \{0,1\}^N$ *is a mask applied to the inputs of* $\mathcal{L}$*, then there is a unique corresponding mask* $\mathbf{b} \in \{0,1\}^N$ *applied to the outputs, i.e., there is a mask* $\mathbf{b}$ *such that for all* $\mathbf{x} \in \{0,1\}^N$*,* $\mathbf{a} \bullet \mathbf{x} = \mathbf{b} \bullet (\mathcal{L}\mathbf{x})$*. The relationship between* $\mathbf{a}$ *and* $\mathbf{b}$ *is given by* $\mathbf{a} = \mathcal{L}'\mathbf{b}$*, where* $\mathcal{L}'$ *is the matrix transpose of* $\mathcal{L}$*.*

*If* $\Delta\mathbf{x} \in \{0,1\}^N$ *is an input difference for* $\mathcal{L}$*, then* $\Delta\mathbf{y} = \mathcal{L}(\Delta\mathbf{x})$ *is the unique corresponding output difference, i.e.,* $\mathcal{L}(\mathbf{x}) \oplus \mathcal{L}(\mathbf{x} \oplus \Delta\mathbf{x}) = \Delta\mathbf{y}$ *for all* $\mathbf{x} \in \{0,1\}^N$*.*

It follows from Lemma 2 that if $\mathbf{a}^t / \Delta\mathbf{x}^t$ and $\mathbf{a}^{t+1} / \Delta\mathbf{x}^{t+1}$ are input and output masks / differences for round $t$, then the resulting input and output masks / differences for the *substitution stage* of round $t$ are $\mathbf{a}^t / \Delta\mathbf{x}^t$ and $\mathbf{b}^t = \mathcal{L}'\mathbf{a}^{t+1} / \Delta\mathbf{y}^t = \mathcal{L}^{-1}(\Delta\mathbf{x}^{t+1})$. Further, $\mathbf{a}^t / \Delta\mathbf{x}^t$ and $\mathbf{b}^t / \Delta\mathbf{y}^t$ can be naturally partitioned into input and output masks / differences for each s-box in round $t$. Enumerate the s-boxes from left to right as $S_1^t, S_2^t, \ldots, S_M^t$, and let the input and output masks / differences for $S_m^t$ be denoted $\mathbf{a}_m^t / \Delta\mathbf{x}_m^t$ and $\mathbf{b}_m^t / \Delta\mathbf{y}_m^t$ $(1 \leq m \leq M)$. Then from Matsui's Piling-up Lemma [16],

$$ELP^t(\mathbf{a}^t, \mathbf{a}^{t+1}) = \prod_{m=1}^{M} LP^{S_m^t}(\mathbf{a}_m^t, \mathbf{b}_m^t) \tag{7}$$

$$EDP^t(\Delta\mathbf{x}^t, \Delta\mathbf{x}^{t+1}) = \prod_{m=1}^{M} DP^{S_m^t}(\Delta\mathbf{x}_m^t, \Delta\mathbf{y}_m^t) \tag{8}$$

(here the superscript $S_m^t$ has the obvious meaning).

**Definition 4 ([25]).** *Let* $\Omega$ *be a* $T$*-round linear / differential characteristic for rounds* $1 \ldots T$*. Then* $\Omega$ *is called* consistent *if, for each s-box in rounds* $1 \ldots T$*, the input and output masks / differences determined by* $\Omega$ *for that s-box are either both zero or both nonzero.*

**Definition 5 ([1]).** *Given a consistent linear / differential characteristic, any s-box for which the resulting input and output masks / differences are nonzero is called* linearly / differentially active *(or just active, when the context is clear).*

For the remainder of this paper, we limit our consideration to consistent characteristics.

**Definition 6.** *Given a linear / differential characteristic, let* $\mathbf{v} \in \{0,1\}^N$ *be the input or output mask / difference for the substitution stage of round* $t$*. Then the active s-boxes in round* $t$ *can be determined from* $\mathbf{v}$ *(without knowing the corresponding output or input mask / difference). We define* $\gamma_{\mathbf{v}}$ *to be the* $M$*-bit vector that encodes this pattern of active s-boxes:* $\gamma_{\mathbf{v}} = \gamma_1\gamma_2\ldots\gamma_M$*, where* $\gamma_i = 1$ *if the* $i^{\text{th}}$ *s-box is active, and* $\gamma_i = 0$ *otherwise, for* $1 \leq i \leq M$*.*

**Definition 7.** *Let* $\gamma, \hat{\gamma} \in \{0,1\}^M$. *Then*

$$W_l[\gamma, \hat{\gamma}] = \# \left\{ \mathbf{y} \in \{0,1\}^N : \gamma_{\mathbf{x}} = \gamma, \ \gamma_{\mathbf{y}} = \hat{\gamma}, \ \text{where } \mathbf{x} = \mathcal{L}'\mathbf{y} \right\}$$
$$W_d[\gamma, \hat{\gamma}] = \# \left\{ \Delta\mathbf{x} \in \{0,1\}^N : \gamma_{\Delta\mathbf{x}} = \gamma, \ \gamma_{\Delta\mathbf{y}} = \hat{\gamma}, \ \text{where } \Delta\mathbf{y} = \mathcal{L}(\Delta\mathbf{x}) \right\}.$$

*Remark 3.* Informally, the value $W_l[\gamma, \hat{\gamma}]$ / $W_d[\gamma, \hat{\gamma}]$ represents the number of ways the linear transformation can "connect" a pattern of active s-boxes in one round ($\gamma$) to a pattern of active s-boxes in the next round ($\hat{\gamma}$).

The diffusive power of a linear transformation is its ability to force some minimum number of s-boxes to be active over a sequence of rounds. This is quantified in the following definition.

**Definition 8 ([5]).** *The* linear / differential branch number, $\mathcal{B}_l$ / $\mathcal{B}_d$, *of an SPN linear transformation is the minimum number of linearly / differentially active s-boxes in two consecutive rounds for any nonzero characteristic:*

$$\mathcal{B}_l = \min \left\{ wt(\gamma_{\mathbf{x}}) + wt(\gamma_{\mathbf{y}}) : \mathbf{y} \in \{0,1\}^N \setminus \mathbf{0}, \ \mathbf{x} = \mathcal{L}'\mathbf{y} \right\} \ /$$
$$\mathcal{B}_d = \min \left\{ wt(\gamma_{\Delta\mathbf{x}}) + wt(\gamma_{\Delta\mathbf{y}}) : \Delta\mathbf{x} \in \{0,1\}^N \setminus \mathbf{0}, \ \Delta\mathbf{y} = \mathcal{L}(\Delta\mathbf{x}) \right\}.$$

*Remark 4.* It is trivial to show that $2 \leq \mathcal{B}_l, \mathcal{B}_d \leq (M+1)$. Hong et al. [6] prove that $\mathcal{B}_l = (M+1)$ if and only if $\mathcal{B}_d = (M+1)$; in this case, the linear transformation is called *maximally diffusive*.

# 3  General Analysis of 2-Round MELP / MEDP

In this section we analyze the 2-round MELP and MEDP for a general SPN, stating results that will be useful in later sections. We focus primarily on the MELP, as the development for the MEDP is essentially parallel (we point out the significant differences in Section 3.1).

Without loss of generality, assume that the linear transformation is omitted from round 2. Therefore the active s-boxes in round 2 can be determined directly from an output mask for round 2, without applying Lemma 2.

Let $\mathbf{a}, \mathbf{b} \in \{0,1\}^N \setminus \mathbf{0}$ be input and output masks, respectively, for round 1 and round 2, and let $f = wt(\gamma_{\mathbf{a}}), \ell = wt(\gamma_{\mathbf{b}})$. Enumerate the active s-boxes in round 1 as $S_1^1, S_2^1, \ldots, S_f^1$, and enumerate the active s-boxes in round 2 as $S_1^2, S_2^2, \ldots, S_\ell^2$. Let $\alpha_i$ be the input mask for $S_i^1$ (derived from $\mathbf{a}$), for $1 \leq i \leq f$, and let $\beta_j$ be the output mask for $S_j^2$ (derived from $\mathbf{b}$), for $1 \leq j \leq \ell$. The characteristics in $ALH(\mathbf{a}, \mathbf{b})$ have the form $\langle \mathbf{a}, \mathbf{y}, \mathbf{b} \rangle$; enumerate the distinct "middle" masks as $\mathbf{y}_1, \mathbf{y}_2, \ldots, \mathbf{y}_W$, where $W = W_l[\gamma_{\mathbf{a}}, \gamma_{\mathbf{b}}]$. The $\mathbf{y}_w$ are input masks for round 2; denote the corresponding output masks for the *substitution stage* of round 1 as $\mathbf{x}_1, \mathbf{x}_2, \ldots, \mathbf{x}_W$ ($\mathbf{x}_w$ and $\mathbf{y}_w$ are related as in the first part of Lemma 2). For a given $\mathbf{x}_w$ ($1 \leq w \leq W$), let $\chi_{(w,i)}$ be the output mask for $S_i^1$ ($1 \leq i \leq f$), and for the corresponding $\mathbf{y}_w$, let $\boldsymbol{v}_{(w,j)}$ be the input mask for $S_j^2$ $(1 \leq j \leq \ell)$. It follows from Theorem 1, Definition 2, and (7) that

$$ELP^{[1\ldots2]}(\mathbf{a},\mathbf{b}) \;=\; \sum_{w=1}^{W} \left( \prod_{i=1}^{f} LP^{S_i^1}(\boldsymbol{\alpha}_i, \boldsymbol{\chi}_{(w,i)}) \cdot \prod_{j=1}^{\ell} LP^{S_j^2}(\boldsymbol{v}_{(w,j)}, \boldsymbol{\beta}_j) \right). \quad (9)$$

It is useful to consider the set of vectors (of length $f + \ell$) of the form

$$V_w = \left\langle \boldsymbol{\chi}_{(w,1)}, \; \boldsymbol{\chi}_{(w,2)}, \; \ldots, \; \boldsymbol{\chi}_{(w,f)}, \; \boldsymbol{v}_{(w,1)}, \; \boldsymbol{v}_{(w,2)}, \; \ldots, \; \boldsymbol{v}_{(w,\ell)} \right\rangle, \quad (10)$$

for $1 \leq w \leq W$. Each coordinate of $V_w$ is an element of $\{0,1\}^n \setminus \mathbf{0}$ (recall that $n$ is the s-box input/output size).

**Lemma 3.** *Given $\mathbf{a}, \mathbf{b} \in \{0,1\}^N \setminus \mathbf{0}$ that satisfy $wt(\gamma_{\mathbf{a}}) + wt(\gamma_{\mathbf{b}}) = \mathcal{B}_l$, let $W = W_l[\gamma_{\mathbf{a}}, \gamma_{\mathbf{b}}]$, $f = wt(\gamma_{\mathbf{a}})$, $\ell = wt(\gamma_{\mathbf{b}})$, and let $\boldsymbol{\chi}_{(w,i)}, \boldsymbol{v}_{(w,j)}$ be defined as above. Then for fixed $i$ $(1 \leq i \leq f)$, the values $\boldsymbol{\chi}_{(1,i)}, \ldots, \boldsymbol{\chi}_{(W,i)}$ are distinct, and for fixed $j$ $(1 \leq j \leq \ell)$, the values $\boldsymbol{v}_{(1,j)}, \ldots, \boldsymbol{v}_{(W,j)}$ are distinct. In other words, for the set of vectors $\{V_w\}_{w=1}^{W}$, all the values in any one position are distinct.*

*Proof.* Contained in the proof of Theorem 1 in [23].

*Remark 5.* Clearly if $wt(\gamma_{\mathbf{a}}) + wt(\gamma_{\mathbf{b}}) = \mathcal{B}_l$, then $W_l[\gamma_{\mathbf{a}}, \gamma_{\mathbf{b}}] \leq (2^n - 1)$. Further, the values $\boldsymbol{\chi}_{(w,i)}$ and $\boldsymbol{v}_{(w,j)}$ depend only on $\gamma_{\mathbf{a}}$ and $\gamma_{\mathbf{b}}$, not on the specific values of $\mathbf{a}$ and $\mathbf{b}$.

**Lemma 4.** *Given $\mathbf{a}, \mathbf{b} \in \{0,1\}^N \setminus \mathbf{0}$ that satisfy $wt(\gamma_{\mathbf{a}}) + wt(\gamma_{\mathbf{b}}) > \mathcal{B}_l$, let $W = W_l[\gamma_{\mathbf{a}}, \gamma_{\mathbf{b}}]$, $f = wt(\gamma_{\mathbf{a}})$, $\ell = wt(\gamma_{\mathbf{b}})$, and let $\boldsymbol{\chi}_{(w,i)}, \boldsymbol{v}_{(w,j)}$ be defined as above. Consider the vectors $V_w$ in (10). Select any $(f + \ell - \mathcal{B}_l)$ vector positions, and fix a value in $\{0,1\}^n \setminus \mathbf{0}$ for each position. Now form the subset of $\{V_w\}_{w=1}^{W}$ consisting of those $V_w$ that contain the selected fixed values in the specified positions—denote this subset by $\mathcal{V}$. Then for each of the $\mathcal{B}_l$ vector positions whose values were not fixed, all the values in that position are distinct as we range over $\mathcal{V}$.*

*Proof.* Contained in the proof of Theorem 1 in [23].

*Remark 6.* It follows that the number of vectors in $\mathcal{V}$ is at most $(2^n - 1)$. The vectors in $\mathcal{V}$ depend on $\gamma_{\mathbf{a}}$ and $\gamma_{\mathbf{b}}$ (not on the specific values of $\mathbf{a}$ and $\mathbf{b}$), and also on the choice of vector positions to be assigned fixed values, together with the particular fixed values chosen for those positions.

**Definition 9.** *For $T = 2$ core SPN rounds, a $\mathcal{B}_l$-list is a set of vectors, each of which has length $\mathcal{B}_l$, that has been derived in one of two ways:*

1. *by selecting any $\mathbf{a}, \mathbf{b} \in \{0,1\}^N \setminus \mathbf{0}$ satisfying $wt(\gamma_{\mathbf{a}}) + wt(\gamma_{\mathbf{b}}) = \mathcal{B}_l$ and forming the set $\{V_w\}_{w=1}^{W}$, as in Lemma 3;*

2. *by selecting any* $\mathbf{a}, \mathbf{b} \in \{0,1\}^N \setminus \mathbf{0}$ *satisfying* $wt(\gamma_{\mathbf{a}}) + wt(\gamma_{\mathbf{b}}) > \mathcal{B}_l$, *forming any set* $\mathcal{V}$ *according to Lemma 4, and then shortening each vector in* $\mathcal{V}$ *to length* $\mathcal{B}_l$ *by removing those positions that were assigned fixed values.*

Let $\mathcal{B}_l\text{-}LIST^{(i)}$ denote the set of all $\mathcal{B}_l$-lists that are formed by Option $i$ above, for $i = 1, 2$, and let

$$\mathcal{B}_l\text{-}LIST = \mathcal{B}_l\text{-}LIST^{(1)} \cup \mathcal{B}_l\text{-}LIST^{(2)} .$$

For any $\mathcal{Z} \in \mathcal{B}_l\text{-}LIST$, let $\delta(\mathcal{Z})$ denote the number of vectors in $\mathcal{Z}$.

It follows from Remarks 5 and 6 that $\delta(\mathcal{Z}) \leq (2^n - 1)$ for any $\mathcal{Z} \in \mathcal{B}_l\text{-}LIST$. For any vector $\mathbf{z} = \langle \zeta_1, \zeta_2, \ldots, \zeta_{\mathcal{B}_l} \rangle$ in any $\mathcal{B}_l$-list $\mathcal{Z}$, each $\zeta_i$ is either an output mask for a particular s-box in round 1, or an input mask for a particular s-box in round 2. In the former case, let $\alpha_i$ denote a nonzero *input* mask for the same s-box, and let $LP^*(\alpha_i, \zeta_i) = LP(\alpha_i, \zeta_i)$. In the latter case, let $\alpha_i$ denote a nonzero *output* mask for the same s-box, and let $LP^*(\alpha_i, \zeta_i) = LP(\zeta_i, \alpha_i)$; here $\alpha_i$ is playing the role of one of the $\beta_j$ used earlier, e.g., in (9), and $LP^*(\cdot, \cdot)$ is the *transpose* of the s-box LP table. (For simplicity, the specific s-box is now implicit in the notation.)

**Definition 10.** *Let* $\mathcal{Z} \in \mathcal{B}_l\text{-}LIST$. *Then*

$$\sigma(\mathcal{Z}) \overset{def}{=} \max_{\alpha_1, \ldots, \alpha_{\mathcal{B}_l} \in \{0,1\}^n \setminus \mathbf{0}} \left( \sum_{\langle \zeta_1, \ldots, \zeta_{\mathcal{B}_l} \rangle \in \mathcal{Z}} \prod_{i=1}^{\mathcal{B}_l} LP^*(\alpha_i, \zeta_i) \right) .$$

The following two theorems are central to this paper.

**Theorem 2.** *The 2-round MELP is lower bounded by*

$$\max \left\{ \sigma(\mathcal{Z}) : \mathcal{Z} \in \mathcal{B}_l\text{-}LIST^{(1)} \right\} .$$

*Proof.* It is easy to see that $\max \left\{ \sigma(\mathcal{Z}) : \mathcal{Z} \in \mathcal{B}_l\text{-}LIST^{(1)} \right\}$ is exactly equal to

$$\max_{\substack{\mathbf{a}, \mathbf{b} \in \{0,1\}^N \setminus \mathbf{0} \\ wt(\gamma_{\mathbf{a}}) + wt(\gamma_{\mathbf{b}}) = \mathcal{B}_l}} ELP^{[1 \ldots 2]}(\mathbf{a}, \mathbf{b}) ,$$

which clearly lower bounds the 2-round MELP (see (3)).

**Theorem 3.** *The 2-round MELP is upper bounded by*

$$\max \{ \sigma(\mathcal{Z}) : \mathcal{Z} \in \mathcal{B}_l\text{-}LIST \} .$$

*Proof.* Let $\mathcal{M} = \max \{ \sigma(\mathcal{Z}) : \mathcal{Z} \in \mathcal{B}_l\text{-}LIST \}$. Given the proof of Theorem 2, it suffices to show that

$$\max_{\substack{\mathbf{a}, \mathbf{b} \in \{0,1\}^N \setminus \mathbf{0} \\ wt(\gamma_{\mathbf{a}}) + wt(\gamma_{\mathbf{b}}) > \mathcal{B}_l}} ELP^{[1 \ldots 2]}(\mathbf{a}, \mathbf{b}) \leq \mathcal{M} .$$

Let $\mathbf{a}, \mathbf{b} \in \{0,1\}^N \setminus \mathbf{0}$ such that $F \stackrel{\text{def}}{=} wt(\gamma_{\mathbf{a}}) + wt(\gamma_{\mathbf{b}}) - \mathcal{B}_l > 0$. In keeping with Lemma 4, isolate $F$ of the active s-boxes to be assigned fixed output/input masks (fixed output masks for round-1 s-boxes, and fixed input masks for round-2 s-boxes), let these fixed masks be denoted $\overline{\zeta}_1, \ldots, \overline{\zeta}_F$, and let the corresponding input/output masks derived from $\mathbf{a}$ and $\mathbf{b}$ be denoted $\overline{\alpha}_1, \ldots, \overline{\alpha}_F$. Let the masks derived from $\mathbf{a}$ and $\mathbf{b}$ for the "non-fixed" s-boxes be denoted $\alpha_1, \ldots, \alpha_{\mathcal{B}_l}$. Denote the $\mathcal{B}_l$-list resulting from this setup by $\mathcal{Z}_{\overline{\zeta}_1, \ldots, \overline{\zeta}_F}$. Then

$$ELP^{[1\ldots2]}(\mathbf{a}, \mathbf{b})$$

$$= \sum_{\overline{\zeta}_1, \ldots, \overline{\zeta}_F \in \{0,1\}^n \setminus \mathbf{0}} \ \sum_{\langle \zeta_1, \ldots, \zeta_{\mathcal{B}_l} \rangle \in \mathcal{Z}_{\overline{\zeta}_1, \ldots, \overline{\zeta}_F}} \left( \prod_{i=1}^{F} LP^*(\overline{\alpha}_i, \overline{\zeta}_i) \cdot \prod_{j=1}^{\mathcal{B}_l} LP^*(\alpha_j, \zeta_j) \right)$$

$$= \sum_{\overline{\zeta}_1, \ldots, \overline{\zeta}_F \in \{0,1\}^n \setminus \mathbf{0}} \prod_{i=1}^{F} LP^*(\overline{\alpha}_i, \overline{\zeta}_i) \left( \sum_{\langle \zeta_1, \ldots, \zeta_{\mathcal{B}_l} \rangle \in \mathcal{Z}_{\overline{\zeta}_1, \ldots, \overline{\zeta}_F}} \prod_{j=1}^{\mathcal{B}_l} LP^*(\alpha_j, \zeta_j) \right)$$

$$\leq \mathcal{M} \left( \sum_{\overline{\zeta}_1, \ldots, \overline{\zeta}_F \in \{0,1\}^n \setminus \mathbf{0}} \prod_{i=1}^{F} LP^*(\overline{\alpha}_i, \overline{\zeta}_i) \right)$$

$$= \mathcal{M},$$

where the last equality follows easily from (1).

## 3.1   Considerations Specific to MEDP

Tailoring the above definitions and results to the MEDP is straightforward: $\mathcal{B}_d$ is substituted for $\mathcal{B}_l$, $W_d[\ ]$ for $W_l[\ ]$, "difference" for "mask," DP values for LP values, and $DIFF(\cdot, \cdot)$ for $ALH(\cdot, \cdot)$. As well, the relationship between input and output differences over the linear transformation is via the second part of Lemma 2.

## 4   Lower Bounding the AES 2-Round MELP / MEDP

For the AES, $\mathcal{B}_l = \mathcal{B}_d = 5$; this is due to the fact that $\mathcal{B}_l = \mathcal{B}_d = 5$ for the 32-bit linear transformation component of the 128-bit AES linear transformation (see Figure 1) [5]. Hereafter we refer to $\mathcal{B}_l$-lists or $\mathcal{B}_d$-lists as 5-lists. As noted earlier, all AES s-boxes are identical. It is not hard to see that computing the MELP / MEDP for 2 AES rounds is equivalent to computing the MELP / MEDP for the "reduced" SPN depicted in Figure 2.

To lower bound the 2-round MELP / MEDP for the AES, we compute the value in Theorem 2 (or its MEDP counterpart) for the SPN in Figure 2. There are 56 pairs $(\gamma, \hat{\gamma}) \in \{0,1\}^4 \times \{0,1\}^4$ for which $wt(\gamma) + wt(\hat{\gamma}) = 5$; enumerate these as $(\gamma_1, \hat{\gamma}_1), (\gamma_2, \hat{\gamma}_2), \ldots, (\gamma_{56}, \hat{\gamma}_{56})$. A straightforward computation reveals that the 5-list associated with each pair $(\gamma_s, \hat{\gamma}_s)$ contains exactly $(2^8 - 1) = 255$

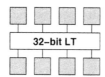

**Fig. 2.** Reduced 2-round AES

| | | |
|---|---|---|
| 1. | | LowerBound = 0 |
| 2. | | For $s = 1$ to 56 |
| 3. | | For $1 \leq \alpha_1, \alpha_2, \alpha_3, \alpha_4, \alpha_5 \leq 255$ |
| 4. | | Sum = 0 |
| 5. | | For $w = 1$ to 255 |
| 6. | | Prod = 1 |
| 7. | | For $i = 1$ to 5 |
| 8. | | Prod = Prod $\times$ $XP^*(\alpha_i, D[s, w, i])$ |
| 9. | | End For |
| 10. | | Sum = Sum + Prod |
| 11. | | End For |
| 12. | | If (Sum > LowerBound) |
| 13. | | LowerBound = Sum |
| 14. | | End If |
| 15. | | End For |
| 16. | End For | |

**Fig. 3.** Algorithm for lower bounding the 2-round MELP / MEDP for the AES

vectors. We store all the 5-lists in a 3-dimensional array of bytes, $D[\cdot, \cdot, \cdot]$, of size $56 \times 255 \times 5$, such that $D[s, \cdot, \cdot]$ contains the 5-list for $(\gamma_s, \hat{\gamma}_s)$ in the obvious way. Computing the lower bound on the MELP / MEDP reduces to the algorithm in Figure 3. (We use XP to mean either LP or DP, as appropriate.) The algorithm as presented is computationally intensive; however, we can make use of the fact that we are searching for a maximum to incorporate significant pruning, greatly reducing the running time.

Using the above algorithm, the lower bound on the 2-round MELP for the AES is $1.638 \times 2^{-28}$. Since the best upper bound is $\frac{48,193,441}{2^{52}} \approx 1.44 \times 2^{-27}$ [3, 23], the 2-round MELP is now known almost exactly. The lower bound on the 2-round MEDP is $1.656 \times 2^{-29}$, and since the best upper bound is $\frac{79}{2^{34}} \approx 1.23 \times 2^{-28}$ [3, 23], the 2-round MEDP is also now known almost exactly.

These lower bounds are important in light of the fact, stated earlier, that the 4th power of any upper bound on the 2-round MELP / MEDP for the AES (including the exact value) is an upper bound on the MELP / MEDP for $T \geq 4$ [23, 24]. This is how Park et al. obtain the upper bounds in the last row of Table 1. However, by constraining the 2-round MELP / MEDP as above, we see that this approach is essentially exhausted (for the AES).

## 5    Best AES 2-Round Upper Bounds Not Tight

In this section we show that the best upper bounds on the 2-round MELP and MEDP for the AES are not tight. First, we state the rationale behind the current bounds, and then we explain why they cannot be attained.

It is well known that all the nontrivial rows and columns of the LP / DP table for the AES s-box have the same distribution of values, given in Table 2 for the LP table and Table 3 for the DP table ($\rho_i$ is a distinct value, and $\phi_i$ is the frequency with which it occurs) [11, 23].

**Table 2.** Distribution of LP values for the AES s-box

| $i$ | 1 | 2 | 3 | 4 | 5 | 6 | 7 | 8 | 9 |
|---|---|---|---|---|---|---|---|---|---|
| $\rho_i$ | $\left(\frac{8}{64}\right)^2$ | $\left(\frac{7}{64}\right)^2$ | $\left(\frac{6}{64}\right)^2$ | $\left(\frac{5}{64}\right)^2$ | $\left(\frac{4}{64}\right)^2$ | $\left(\frac{3}{64}\right)^2$ | $\left(\frac{2}{64}\right)^2$ | $\left(\frac{1}{64}\right)^2$ | 0 |
| $\phi_i$ | 5 | 16 | 36 | 24 | 34 | 40 | 36 | 48 | 17 |

**Table 3.** Distribution of DP values for the AES s-box

| $i$ | 1 | 2 | 3 |
|---|---|---|---|
| $\rho_i$ | $\frac{1}{64}$ | $\frac{1}{128}$ | 0 |
| $\phi_i$ | 1 | 126 | 129 |

We again use XP to mean either LP or DP, as appropriate. Consider the upper bound given in Theorem 3 (or its MEDP counterpart). Let $\mathcal{Z} \in$ 5-LIST. Adapting Theorem 1 in [23] to our notation, $\sigma(\mathcal{Z})$ equals the maximum value possible for a 5-list if, for some $\alpha_1, \ldots, \alpha_5 \in \{0, 1\}^8 \setminus \mathbf{0}$, the following two conditions are satisfied:

1. For every $\langle \zeta_1, \ldots, \zeta_5 \rangle \in \mathcal{Z}$,

$$XP^*(\alpha_1, \zeta_1) = XP^*(\alpha_2, \zeta_2) = \cdots = XP^*(\alpha_5, \zeta_5).$$

2. No nonzero XP value is omitted. In other words, for each $i \in \{1 \ldots 5\}$, as we range over all $\langle \zeta_1, \ldots, \zeta_5 \rangle \in \mathcal{Z}$ the values $XP^*(\alpha_i, \zeta_i)$ include every nonzero value in the XP table row or column indexed by $\alpha_i$. (Obviously a necessary condition for the omission of a nonzero XP value is that $\delta(\mathcal{Z}) < 255$.)

Using the notation of Table 2 / Table 3, the maximum possible value for $\sigma(\mathcal{Z})$ is $\sum \rho_i^5 \phi_i$ — this is exactly the best upper bound on the 2-round MELP /

MEDP [3, 23]. However, we have determined that this "worst-case" situation never occurs.

In brief, we systematically generated all 5-lists in 5-LIST$^{(2)}$ (it is clear from Section 4 that we don't need to consider the elements of 5-LIST$^{(1)}$). We observed that $\delta(\mathcal{Z}) < 255$ for all $\mathcal{Z} \in$ 5-LIST$^{(2)}$ (specifically, $251 \leq \delta(\mathcal{Z}) \leq 254$). For each 5-list generated, we ran an algorithm which ascertained that there do not exist masks $\alpha_1, \ldots, \alpha_5 \in \{0, 1\}^8 \setminus \mathbf{0}$ satisfying both Condition 1 and Condition 2 above. We used aggressive pruning to avoid iterating through all possible values of the $\alpha_i$ (approximately $2^{40}$) for any 5-list.

## 6    Modified Version of KMT2 Algorithm

The KMT2 algorithm (resp. its dual, KMT2-DC), due to Keliher, Meijer, and Tavares [8, 11], is a general algorithm that can be used to compute an upper bound on the MELP (resp. MEDP) for each value of $T \geq 2$ for any SPN. For a fixed value $T \geq 2$, and for all nonzero patterns of active s-boxes in the first and last rounds given by $\gamma$ and $\hat{\gamma}$, KMT2 (resp. KMT2-DC) computes a value $UB^{[1 \ldots T]}(\gamma, \hat{\gamma})$ such that for all $\mathbf{a}, \mathbf{b} \in \{0, 1\}^N \setminus \mathbf{0}$, if $\gamma_{\mathbf{a}} = \gamma$ and $\gamma_{\mathbf{b}} = \hat{\gamma}$, then $ELP^{[1 \ldots T]}(\mathbf{a}, \mathbf{b}) \leq UB^{[1 \ldots T]}(\gamma, \hat{\gamma})$ (resp. $EDP^{[1 \ldots T]}(\mathbf{a}, \mathbf{b}) \leq UB^{[1 \ldots T]}(\gamma, \hat{\gamma})$). The values in the third row of Table 1 are from KMT2 and KMT2-DC.

The KMT2 / KMT2-DC algorithm works recursively, i.e., for $T \geq 3$, the values $UB^{[1 \ldots T]}(\gamma, \hat{\gamma})$ depend on the values $UB^{[1 \ldots (T-1)]}(\gamma, \hat{\gamma})$. This allows for a very simple improvement: For any $T \geq 2$, suppose that $B$ is known to be an upper bound on the MELP / MEDP for that value of $T$ (from some external source of information). Then if $B < UB^{[1 \ldots T]}(\gamma, \hat{\gamma})$, replace $UB^{[1 \ldots T]}(\gamma, \hat{\gamma})$ with $B$ before proceeding to the computations for $T + 1$. In other words, enhance

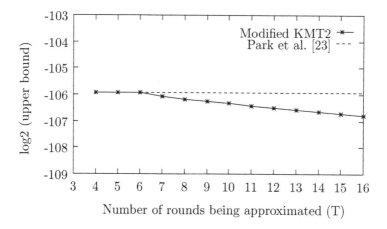

**Fig. 4.** Results from modified KMT2 for AES

KMT2 / KMT2-DC by incorporating other upper bounds when those bounds are superior to the values determined directly by the algorithm.

For the AES, we modified KMT2 in this fashion by incorporating the upper bound on the MELP for $T \geq 4$ due to Park et al [23]. The results are plotted in Figure 4. For $T \geq 8$, for example, the upper bound on the MELP is improved to $1.778 \times 2^{-107}$. This improvement is slight, but it is an effective "proof of concept." For other ciphers, the modified version of KMT2 / KMT2-DC may yield much more significant improvements over existing upper bounds.

Interestingly, modifying KMT2-DC using the upper bound on the AES MEDP for $T \geq 4$ due to Park et al. yielded no improvement over the existing bound for $T \geq 4$. This appears to be an artifact of the simple distribution of DP values for the AES s-box (LP / DP values play a fundamental role in KMT2 / KMT2-DC).

## 7   Conclusion

We have carefully analyzed bounds related to linear and differential cryptanalysis for the AES. We present nontrivial lower bounds on the 2-round maximum expected linear probability (MELP) and maximum expected differential probability (MEDP), trapping each value in a small interval. We then prove that the best upper bounds on the 2-round MELP and MEDP are not tight. Finally, we show how a modified version of the KMT2 / KMT2-DC algorithm can potentially improve existing upper bounds on the MELP / MEDP for any SPN, and we use the modified KMT2 algorithm to tighten the upper bound on the AES MELP to $1.778 \times 2^{-107}$, for $T \geq 8$.

## References

1. E. Biham, *On Matsui's linear cryptanalysis*, Advances in Cryptology—EUROCRYPT'94, LNCS 950, pp. 341–355, Springer-Verlag, 1995.
2. E. Biham and A. Shamir, *Differential cryptanalysis of DES-like cryptosystems*, Advances in Cryptology—CRYPTO'90, LNCS 537, pp. 2–21, Springer-Verlag, 1991.
3. K. Chun, S. Kim, S. Lee, S.H. Sung, S. Yoon, *Differential and linear cryptanalysis for 2-round SPNs*, Information Processing Letters, Vol. 87, pp. 277–282, 2003.
4. J. Daemen, L. Knudsen, and V. Rijmen, *The block cipher* SQUARE, Fast Software Encryption (FSE'97), LNCS 1267, pp. 149–165, Springer-Verlag, 1997.
5. J. Daemen and V. Rijmen, *The Design of Rijndael: AES—The Advanced Encryption Standard*, Springer-Verlag, 2002.
6. S. Hong, S. Lee, J. Lim, J. Sung, and D. Cheon, *Provable security against differential and linear cryptanalysis for the SPN structure*, Fast Software Encryption (FSE 2000), LNCS 1978, pp. 273–283, Springer-Verlag, 2001.
7. J.-S. Kang, S. Hong, S. Lee, O. Yi, C. Park, and J. Lim, *Practical and provable security against differential and linear cryptanalysis for substitution-permutation networks*, ETRI Journal, Vol. 23, No. 4, December 2001.
8. L. Keliher, *Linear cryptanalysis of substitution-permutation networks*, Ph.D. Thesis, Queen's University, Kingston, Canada, 2003.

9. L. Keliher, H. Meijer, and S. Tavares, *New method for upper bounding the maximum average linear hull probability for SPNs,* Advances in Cryptology— EUROCRYPT 2001, LNCS 2045, pp. 420–436, Springer-Verlag, 2001.

10. L. Keliher, H. Meijer, and S. Tavares, *Dual of new method for upper bounding the maximum average linear hull probability for SPNs,* Technical Report, IACR ePrint Archive (http://eprint.iacr.org, Paper # 2001/033), 2001.

11. L. Keliher, H. Meijer, and S. Tavares, *Improving the upper bound on the maximum average linear hull probability for Rijndael,* Eighth Annual International Workshop on Selected Areas in Cryptography (SAC 2001), LNCS 2259, pp. 112–128, Springer-Verlag, 2001.

12. L. Keliher, H. Meijer, and S. Tavares, *Completion of computation of improved upper bound on the maximum average linear hull probability for Rijndael,* Technical Report, IACR ePrint Archive (http://eprint.iacr.org, Paper # 2004/074), 2004.

13. L. Knudsen, *Practically secure Feistel ciphers,* Fast Software Encryption, LNCS 809, pp. 211–221, Springer-Verlag, 1994.

14. X. Lai, J. Massey, and S. Murphy, *Markov ciphers and differential cryptanalysis,* Advances in Cryptology—EUROCRYPT'91, LNCS 547, pp. 17–38, Springer-Verlag, 1991.

15. C.H. Lim, *CRYPTON: A new 128-bit block cipher,* The First Advanced Encryption Standard Candidate Conference, Proceedings, Ventura, California, August 1998.

16. M. Matsui, *Linear cryptanalysis method for DES cipher,* Advances in Cryptology— EUROCRYPT'93, LNCS 765, pp. 386–397, Springer-Verlag, 1994.

17. M. Matsui, *On correlation between the order of s-boxes and the strength of DES,* Advances in Cryptology—EUROCRYPT'94, LNCS 950, pp. 366–375, Springer-Verlag, 1995.

18. W. Meier and O. Staffelbach, *Nonlinearity criteria for cryptographic functions,* Advances in Cryptology—EUROCRYPT'89, LNCS 434, pp. 549–562, Springer-Verlag, 1990.

19. K. Nyberg, *Linear approximation of block ciphers,* Advances in Cryptology— EUROCRYPT'94, LNCS 950, pp. 439–444, Springer-Verlag, 1995.

20. K. Nyberg and L. Knudsen, *Provable security against a differential attack,* Journal of Cryptology, Vol. 8, No. 1, pp. 27–37, 1995.

21. K. Ohta, S. Moriai, and K. Aoki, *Improving the search algorithm for the best linear expression,* Advances in Cryptology—CRYPTO'95, LNCS 963, pp. 157–170, Springer-Verlag, 1995.

22. S. Park, S.H. Sung, S. Chee, E-J. Yoon, and J. Lim, *On the security of Rijndael-like structures against differential and linear cryptanalysis,* Advances in Cryptology— ASIACRYPT 2002, LNCS 2501, pp. 176–191, Springer-Verlag, 2002.

23. S. Park, S.H. Sung, S. Lee, J. Lim, *Improving the upper bound on the maximum differential and the maximum linear hull probability for SPN structures and AES,* Fast Software Encryption (FSE 2003), LNCS 2887, pp. 247–260, Springer-Verlag, 2003.

24. F. Sano, K. Ohkuma, H. Shimizu, and S. Kawamura, *On the security of nested SPN cipher against the differential and linear cryptanalysis,* IEICE Transactions on Fundamentals of Electronics, Communications and Computer Sciences, Vol. E86-A, No. 1, pp. 37–46, 2003.

25. S. Vaudenay, *On the security of CS-Cipher,* Fast Software Encryption (FSE'99), LNCS 1636, pp. 260–274, Springer-Verlag, 1999.

# Some Algebraic Aspects of the Advanced Encryption Standard

Carlos Cid

Information Security Group,
Royal Holloway, University of London,
Egham, Surrey, TW20 0EX, UK
carlos.cid@rhul.ac.uk

**Abstract.** Since being officially selected as the new Advanced Encryption Standard (AES), Rijndael has continued to receive great attention and has had its security continuously evaluated by the cryptographic community.

Rijndael is a cipher with a simple, elegant and *highly* algebraic structure. Its selection as the AES has led to a growing interest in the study of algebraic properties of block ciphers, and in particular algebraic techniques that can be used in their cryptanalysis.

In these notes we will examine some algebraic aspects of the AES and consider a number of algebraic techniques that could be used in the analysis of the cipher. In particular, we will focus on the large, though surprisingly simple, systems of multivariate quadratic equations derived from the encryption operation, and consider some approaches that could be used when attempting to solve these systems.

These notes refer to an invited talk given at the Fourth Conference on the Advanced Encryption Standard (AES4) in May 2004, and are largely based on [4].

## 1 Introduction

Rijndael is a block-cipher with a simple and elegant structure. It has been designed to offer strong resistance against *known attacks*, in particular differential and linear cryptanalysis, while enabling efficient implementation on different platforms. Given its careful design criteria, it seems unlikely that its security can be affected by conventional methods of cryptanalysis.

Rijndael has also a highly algebraic structure: the cipher round transformations are based on simple operations over the finite field $\mathbb{F}_{2^8}$. Its selection as the AES has therefore led to a growing interest in the study of algebraic properties of block ciphers, as well as algebraic techniques that can be used in their cryptanalysis [1, 2, 6, 12, 13].

This new approach in cryptanalysis seems promising. One reason is that conventional methods of cryptanalysis of block-ciphers (e.g. differential and linear cryptanalysis) are generally based on a "statistical" approach: the attacker attempts to construct probabilistic characteristics through as many rounds of the

H. Dobbertin, V. Rijmen, A. Sowa (Eds.): AES 2004, LNCS 3373, pp. 58–66, 2005.

cipher as possible, in order to distinguish the cipher from a random permutation. Most modern ciphers have been designed with these attacks in mind, and therefore do not generally have their security affected by them.

In contrast, the so-called *algebraic attacks* exploit the intrinsic algebraic structure of a cipher: the attacker expresses the encryption transformation as a (large) set of multivariate polynomial equations, and subsequently attempts to solve such a system to recover the encryption key. Algebraic attacks could open new perspectives in the cryptanalysis of block ciphers.

In these notes we will examine some algebraic aspects of the AES and consider a number of algebraic techniques that could be used in the analysis of the cipher.

## 2   The Basic Structure of the AES

Rijndael is a key-iterated block cipher, which alternates key-independent round transformations and key addition. In the basic version (considered here), the cipher encrypts 128-bit blocks in 10 rounds, using 128-bit keys. We refer to [9] for a full description of the cipher.

In these notes we will also consider the *Big Encryption System (BES)*, another iterated block cipher which was introduced in [13]. BES operates on 128-byte blocks with 128-byte keys, and has also a very simple algebraic structure: one round of the cipher consists of inversion, matrix multiplication and key addition, all operations over $\mathbb{F}_{2^8}$. We refer to [13] for the full description of the BES cipher.

Both the AES and the BES use a *state vector* of bytes, which is transformed by the basic operations within a round. Furthermore, it is shown in [13] that one can *embed* the AES into the BES, and that way obtain an alternative description of the AES. This relationship between both ciphers may well provide new ways for the cryptanalysis of the AES.

## 3   Algebraic Analysis of the AES

Due to the rich algebraic structure of the AES, there is currently a growing interest in the study of algebraic techniques which could be applied in its cryptanalysis. These are known as *algebraic attacks*. Currently there appears to be two main approaches:

  − Study the system of polynomial equations derived from the cipher;
  − Study the AES underlying algebraic structure.

## 4   Algebraic Attacks

Unlike most conventional methods of cryptanalysis, the so-called *algebraic attacks* attempt to exploit the intrinsic algebraic structure of the cipher. More specifically, the attacker expresses the encryption transformation as a set of multivariate polynomial equations and tries to recover the encryption key by solving the system.

While in theory most modern block ciphers can be fully described by a system of multivariate polynomial equations over a finite field, for the majority of the cases such systems prove to be just too complex for any practical purpose. However, given its algebraic structure, it seems that the AES could be more vulnerable to such approach.

In [6] Courtois and Pieprzyk exhibit a large, sparse and overdefined system of multivariate quadratic equations over $\mathbb{F}_2$ whose solution would recover the AES encryption key. In the same paper they propose a method called *XSL (eXtended Sparse Linearization)*, as an attempt to efficiently solve the system. Around the same time, Murphy and Robshaw [13] showed how to express the AES encryption as a far simpler system of equations over $\mathbb{F}_{2^8}$, which is derived from the BES. If XSL or some of its variants are in fact valid methods, this system should be faster to solve than the original one over $\mathbb{F}_2$, and in theory, could provide an efficient key-recovery attack.

## 5    Potential Attack Techniques

Given the AES and BES algebraic formulations, it is clear that an efficient method for the solution of this type of system of multivariate quadratic equations would provide a key-recovery attack of the AES with potentially very few plaintext-ciphertext pairs. While the problem of solving generic large systems of multivariate equations of degree greater than one over a finite field is known to be NP-complete, it is conceivable that a technique can be developed which exploits the particular algebraic structure of the AES and BES systems. Below we investigate a few approaches which have been proposed for solving such systems.

## 6    Linearization Methods

The method of *linearization* is a well-known technique for solving large systems of multivariate polynomial equations. In this method one considers all monomials in the system as independent variables and tries to solve the system using linear algebra techniques. In order to apply the method, the number of *linearly independent* equations in the system needs to be approximately the same as the number of terms in the system. When this is not the case, a number of techniques have been proposed to generate enough LI equations.

### 6.1    XL Algorithm

In [5] an algorithm for solving systems of multivariate quadratic equations called *XL* (standing for *eXtended Linearization*) is proposed. XL is a simple algorithm: if $A$ is a system of $m$ quadratic equations $f_i$ in $n$ variables over a field $K$, and $D \in \mathbb{N}$, one executes the following steps:

1. **Multiply:** Generate all the products $\prod_{j=1}^{k} x_{i_j} * f_i$ with $k \leq D - 2$;
2. **Linearize:** Consider each monomial of degree $\leq D$ as a new variable and perform Gaussian elimination on the system obtained in step 1;
3. **Solve:** Assume that step 2 yields at least one univariate equation. Solve this equation;
4. **Repeat:** Simplify the equations and repeat the process to find the values of the other variables.

The hope is that after few iterations the algorithm will yield a solution for the system.

In [5] the authors present some estimates for the complexity of the algorithm for random systems with $m \approx n$. In particular, they provide evidence that XL can solve randomly generated *overdefined* systems of polynomial equations in subexponential time.

The XL algorithm (as in the form above) is a reasonably new idea, and its behaviour is not entirely understood yet. When analysing the algorithm, one must examine two key points:

1. Does the algorithm always terminate?
2. Does the algorithm work as predicted?

**Does XL Always Terminate?** By applying well-known commutative algebra techniques (Hilbert Theory), one can show that there are cases for which the algorithm does not terminate [10]. However, when working over finite fields, this problem can be avoided by adding to the system the underlying field equations $x_i^q - x_i = 0$ .

**Does XL Work as Predicted?** Initially it was suggested that XL could solve systems of polynomial equations in subexponential time when the number of equations exceeded the number of variables by a small number. However there has been strong evidence that some of the heuristics used in the original article were too optimistic [3, 10].

This discrepancy arises from the fact that one may often overestimate the number of *linearly independent* equations generated by the algorithm. There has been recently few papers studying the XL [3, 10], and one could say that the algorithm has just started to be better understood now.

In any case, it is widely agreed that application of the XL algorithm against the polynomial system which arises from the AES (either over $\mathbb{F}_{2^8}$ or $\mathbb{F}_2$) does not provide an efficient attack against the cipher.

## 6.2   Variants of XL

Since the introduction of the XL method, a number of variants have been proposed. These attempt to exploit specific properties of the polynomial systems, such as how overdefined the system is, the order of the field, etc. Of particular relevance for the AES is the method proposed in [6] by Courtois and Pieprzyk.

*XSL* is based on the XL method, but uses the sparsity and specific structure of the equations to mount the attack; instead of multiplying the equations by all monomials up to certain degree, in the XSL algorithm the equations are multiplied only by "carefully selected monomials" (we refer to [6] and its earlier version [7] for a full description of the method). While this has the intention to create less new terms when generating new equations, it is not entirely clear the exact criteria used for selecting the monomials.

The system used in [6] to mount the attack has 8000 quadratic equations and 1600 variables, over $\mathbb{F}_2$ (the variables represent the input/output bits). Two attacks are described in [7]: the first one ignores the key schedule and therefore needs 11 known plaintext/ciphertext pairs (for the AES-128); the second attack uses the key schedule, and in theory could be mounted with a single known plaintext/ciphertext pair. In [6] it is claimed that the second XSL attack would have complexity of $\approx 2^{230}$ and $\approx 2^{255}$ when applied against the 128-bit and 256-bit AES, respectively. So the XSL attack would represent a (theoretical) successful key-recovery attack against the 256-bit AES.

**XSL Attack on the BES.** As said earlier, the $\mathbb{F}_{2^8}$-system derived from the BES is much simpler than the $\mathbb{F}_2$-system presented in [6]. In particular, it is far sparser. This would strongly suggest that the XSL attack is more suited to the BES system than to the original AES system.

Murphy and Robshaw consider in [13, 14] the consequences of the XSL attack against the BES. Using the estimates given in [6], they conclude that if XSL is in fact a *valid technique*, a key-recovery attack against the AES might be possible with a work effort of about $2^{100}$ encryptions. This would clearly represent a successful attack against the AES-128.

**Accuracy of the XSL Estimates.** The XSL algorithm consists basically of two main steps:

1. The equation generation procedure;
2. The $T'$ method at the end of the algorithm.

The first step corresponds to the multiplication of the initial set of equations by selected monomials. This is done in similar manner of the XL algorithm. The $T'$ method is used at the end, and in theory would allow the method to effectively solve the system even when the difference between the total number of terms and the number of linearly independent equations is reasonably large.

The main issue when considering XSL attacks (in fact, all the XL-based attacks) against the AES is how accurate the estimates for the number of *linearly independent* equations are. As explained above, there is evidence that some of the heuristics in the original XL paper were too optimistic. In fact, there is even more concern when considering the XSL algorithm. Additionally, it is not clear how effective the $T'$ method is as a last step of the algorithm. The algorithm is an ad-hoc method, based on a number of heuristics arguments, and although this might not invalidate the XSL technique entirely, it makes it harder to formally examine the algorithm and consider whether the XSL attacks described in [6] work *as claimed*.

In fact, we have considered very small versions of BES, with reduced block length and number of rounds, and smaller field. We ran a few simulations with these versions, and it appears that the attacks do not work in the manner predicted in [6]. Again, while this might not invalidate the XSL technique, it could raise doubts on whether the method is generally applicable against the AES and BES systems. It is clear that more research is needed to determine how effective this technique is against the AES.

# 7   Computational Algebra Techniques

Solving multivariate polynomial systems is a typical problem studied in Algebraic Geometry and Commutative Algebra. The classical algorithm for solving this type of problem is the Buchberger algorithm for calculating Gröbner Bases (see [8] for definitions and description of the algorithm). The algorithm generates a basis for the ideal derived from the set of equations, which can then be used to obtain the solutions.

The complexity of most algorithms used for calculating a Gröbner basis of an ideal is closely related to the total degree of the intermediate polynomials that are generated during the running of algorithm. In the worst case the Buchberger algorithm is known to run in double exponential time. One of the most efficient algorithms known, due to Faugère [11], appears to be single exponential. In any case, in practice it is widely believed that Gröbner Bases algorithms cannot be used for efficiently solving *generic* systems with more than a handful of variables.

However, the type of systems which arise from cryptosystems are often very structured. In particular, the BES system has a very regular structure. This is given for $j = 0, \ldots, 15$ and $m = 0, \ldots, 7$ by:

$$
\begin{aligned}
0 &= w_{0,(j,m)} + p_{(j,m)} + k_{0,(j,m)}, & \\
0 &= x_{i,(j,m)} w_{i,(j,m)} + 1 & \text{for } i = 0, \ldots, 9, \\
0 &= w_{i,(j,m)} + k_{i,(j,m)} + \sum_{(j',m')} \alpha_{(j,m),(j',m')} x_{i-1,(j',m')} & \text{for } i = 1, \ldots, 9, \\
0 &= c_{(j,m)} + k_{10,(j,m)} + \sum_{(j',m')} \beta_{(j,m),(j',m')} x_{9,(j',m')}. &
\end{aligned}
$$

This system could be viewed as an "iterated" system of equations, with blocks of similar "sub-systems" repeated for every round. One could also use the transformation $\mathbf{x} \mapsto \mathbf{x}^{254}$ as the S-Box inversion to eliminate a number of variables (the BES system considered above has the simplest form, with only quadratic and linear equations).

Furthermore, since the system includes the equations relating every variable with its conjugates, we have the following easy proposition:

**Proposition 1.** *The degree of polynomials occurring in the computation of a Gröbner basis of a BES-type system with $n$ variables is at most $n$.*

This is clearly an upper bound, and we expect that in practice the degrees are much lower. This fact, together with the particular structure of the system, can be exploited to infer more precise bounds for the complexity of the attack.

One can also seek alternatives for the use of the usual Gröbner bases algorithm. There are also a number of common techniques used in cryptanalysis that could be used in conjunction with computer algebra methods . For example, one could attempt to adapt the *meet-in-the-middle* technique and consider two smaller systems. This has the potential to reduce the complexity of the attack. One should also note that in practice the attacker is not primarily interested in the full solution of the system, but rather in the key variables. In fact, in a "partial key recovery" attack, only few key variables might suffice.

Therefore, it is possible that one may be able to use a combination of cryptanalytic and algebraic techniques (including Linearisation and Gröbner Bases) to mount a successful attack without actually computing the solution of the entire system.

## 7.1    The Polynomial Ideal Generated by the BES System

Let $S$ be the system of multivariate quadratic equations derived from the BES encryption operation, and $K$ a fixed encryption key. A closer look at the properties of the ideal generated by these polynomials may prove to be useful when attempting to solve the system.

For every plaintext/ciphertext pair $(P, C)$, we have a particular system $S_{(P,C)}$ and an ideal [1]

$$I_{(P,C)} = \langle S_{(P,C)} \rangle \subseteq \mathbb{K}[x_{i,(j,m)}, \ldots, w_{i,(j,m)}, \ldots, k_{i,(j,m)}].$$

In fact we are mostly interested in the ideal

$$I^K_{(P,C)} = I_{(P,C)} \cap \mathbb{K}[k_0, k_1, \ldots, k_{15}]$$

where $k_0, k_1, \ldots, k_{15}$ are the first key variables (i.e. the original key).

Thus for every key $K$, we can associate an ideal of $\mathbb{F}[k_0, k_1, \ldots, k_{15}]$ defined as

$$I_K = \bigoplus_{(P,C)} I^K_{(P,C)},$$

where $(P, C)$ run through all plaintext/ciphertext pairs.

Given a key $K$, a random plaintext block $P$, and $C$ such that $E_K(P) = C$, the probability that there exists another key $K'$ with $E_{K'}(P) = C$ is approximately $(1 - 1/(e - 1)) \cong 42\%$. Therefore we expect that in many cases, for a given plaintext/ciphertext pair $(P, C)$, the $\mathbb{K}$-dimension of the residue class ring $\mathbb{K}[k_0, k_1, \ldots, k_{15}]/I^K_{(P,C)}$ is greater than 1 (i.e., the corresponding reduced Gröbner basis should contain polynomials with degree greater than 1).

---

[1] To avoid inconsistent systems, we will make sure to describe the system in such way that it does not include "0-inversions" (i.e. use the map $x \mapsto x^{254}$ when necessary).

On the other hand, the $\mathbb{K}$-dimension of $\mathbb{K}[k_0, k_1, \ldots, k_{15}]/I_K$ is almost certainly 1. In other words, we expect $I_K$ to be of the form

$$I_K = < k_0 - \kappa_0, k_1 - \kappa_1, \ldots, k_{15} - \kappa_{15} >$$

with $\kappa_i \in \mathbb{K}$. If this is not true, then there are *at least* two keys $K_1$ and $K_2$ such that

$$E_{K_1}(P) = E_{K_2}(P)$$

for *every plaintext block* $P$, and $K_1$ and $K_2$ induce the same permutation on the set of possible plaintext blocks, which would not appear to be the case for the AES.

# 8 Alternative Approaches

It is clear that an efficient method for solving the polynomial systems considered so far would represent a successful *key-recovery* attack against the AES. However, even when the system cannot be solved, other approaches could well be used in order to mount less ambitious attacks against the cipher. One could examine common applications of the AES, such as AES-based hash function and MAC constructions, modes of operation, relation between plaintexts, etc.

At the very least, a cryptanalyst would like to find a polynomial-time distinguisher between the cipher and a random permutation. This could be used either to mount a practical attack or simply to show some structural weakness of the cipher.

Given the rich algebraic structure of the cipher, it is not inconceivable that an "algebraic" distinguisher exists. This would most likely exploit the byte-oriented structure of the cipher and the typical round version of the BES, which consists of inversion, matrix multiplication and key addition over $\mathbb{F}_{2^8}$:

$$\mathbf{b} \mapsto M_B.\mathbf{b}^{-1} + (\mathbf{k}_B)_i$$

*Mathematically*, this seems to be the most natural representation of the cipher. Both the S-Box (inversion on $\mathbb{F}_{2^8}$) and the linear layer are highly structured, and this could well be exploited in the analysis of the cipher.

# 9 Conclusion

Rijndael is a cipher with a simple, elegant and *highly* algebraic structure. Its selection as the AES has led to a growing interest in the study of algebraic properties of block ciphers, with a particular focus on algebraic techniques that can be used in their cryptanalysis. One promising approach is to exploit the large, though surprisingly simple, system of multivariate quadratic equations derived from the cipher. An efficient method for solving this system would represent a successful key-recovery attack against the AES. While the problem of solving such systems is known to be hard, it is not entirely unlikely that a technique can be developed which exploits the particular algebraic structure of these particular systems.

Furthermore, it is also possible that the AES algebraic structure could be exploited on mounting less ambitious attacks. The AES has a rich algebraic structure, and while many of these properties might not prove to be relevant in the cryptanalysis, it is not inconceivable that one could find a novel way to explore this structure in the analysis of the cipher.

# References

1. Elad Barkan and Eli Biham. In how many ways can you write Rijndael? In Yuliang Zheng, editor, *Advances in Cryptology - ASIACRYPT 2002*, volume 2501 of *Lecture Notes in Computer Science*, pages 160–175. Springer, 2002.
2. Alex Biryukov and Christophe De Canniere. Block Ciphers and Systems of Quadratic Equations. In *FSE'2003*, 2003.
3. Jiun-Ming Chen and Bo-Yin Yang. Theoretical Analysis of XL over Small Fields. In *Proceedings of the 9th Australasian Conference on Information Security and Privacy*, 2004. to appear.
4. Carlos Cid, Sean Murphy, and Matthew Robshaw. Computational and Algebraic Aspects of the Advanced Encryption Standard. In *Proceedings of the Seventh International Workshop on Computer Algebra in Scientific Computing - CASC 2004*, 2004. to appear.
5. Nicolas Courtois, Alexander Klimov, Jacques Patarin, and Adi Shamir. Efficient Algorithms for Solving Overdefined Systems of Multivariate Polynomial Equations. In *Eurocrypt'2000*, pages 392–407. Springer, 2000.
6. Nicolas Courtois and Josef Pieprzyk. Cryptanalysis of Block Ciphers with Overdefined Systems of Equations. In Yuliang Zheng, editor, *Advances in Cryptology - ASIACRYPT 2002*, volume 2501 of *Lecture Notes in Computer Science*, pages 267–287. Springer, 2002.
7. Nicolas Courtois and Josef Pieprzyk. Cryptanalysis of Block Ciphers with Overdefined Systems of Equations. Cryptology ePrint Archive, Report 2002/044, 2002.
8. David Cox, John Little, and Donal O'Shea. *Ideals, Varieties, and Algorithms*. Undergraduate Texts in Mathematics. Springer, Second edition, 1997.
9. Joan Daemen and Vincent Rijmen. *The Design of Rijndael*. Springer-Verlag, 2002.
10. Claus Diem. The XL-algorithm and a conjecture from commutative algebra, 2004. submitted.
11. Jean-Charles Faugère. A new efficient algorithm for computing Gröbner Bases without reduction to zero F5. In T. Mora, editor, *International Symposium on Symbolic and Algebraic Computation - ISSAC 2002*, pages 75–83, July 2002.
12. N. Ferguson, R. Shroeppel, and D. Whiting. A simple algebraic representation of Rijndael. In *Proceedings of Selected Areas in Cryptography*, pages 103–111. Springer-Verlag, 2001.
13. Sean Murphy and Matthew Robshaw. Essential Algebraic Structure within the AES. In M. Yung, editor, *Advances in Cryptology - CRYPTO 2002*, volume 2442 of *LNCS*, pages 1–16. Springer-Verlag, 2002.
14. Sean Murphy and Matthew Robshaw. Comments on the Security of the AES and the XSL Technique. *Electronic Letters*, 39:26–38, 2003.

# General Principles of Algebraic Attacks and New Design Criteria for Cipher Components*

Nicolas T. Courtois

Axalto Cryptographic Research & Advanced Security,
36-38 rue de la Princesse, BP 45, 78430 Louveciennes Cedex, France
courtois@minrank.org
http://www.nicolascourtois.net

**Abstract.** This paper is about the design of multivariate public key schemes, as well as block and stream ciphers, in relation to recent attacks that exploit various types of multivariate algebraic relations. We survey these attacks focusing on their common fundamental principles and on how to avoid them. From this we derive new very general design criteria, applicable for very different cryptographic components. These amount to avoiding (if possible) the existence of, in some sense "too simple" algebraic relations. Though many ciphers that do not satisfy this new paradigm probably still remain secure, the design of ciphers will never be the same again.

**Keywords:** algebraic attacks, polynomial relations, multivariate equations, finite fields, design of cryptographic primitives, generalised linear cryptanalysis, multivariate public key encryption and signature schemes, HFE, Quartz, Sflash, stream ciphers, Boolean functions, combiners with memory, block ciphers, AES, Rijndael, Serpent, elimination methods, Gröbner bases.

## 1  Introduction

In this paper we consider a very ambitious question: how to design secure cryptosystems and in particular how to design secure ciphers ? Very little real answers do exist in this area. However it is possible to learn from our experience, and formulate some design criteria, resulting on the one hand, from some practical requirements on cryptographic systems, and on the other hand, from the known attacks. Doing so we are still not done, and this for two reasons. First of all, the recommandations do usually conflict with each other and are not obvious to balance. Moreover for both practical implementation criteria and security criteria, it is always hard to know and debatable to what extent exactly a system satisfies these. Nevertheless, the work on the design criteria is and always was an important and necessary area of research.

---

* Work supported by the French Ministry of Research RNRT Project "X-CRYPT".

H. Dobbertin, V. Rijmen, A. Sowa (Eds.): AES 2004, LNCS 3373, pp. 67–83, 2005.

This paper is about an emergence of a new type of design criteria on various types of cryptographic primitives. It turns out that many recent attacks on public key signature and encryption schemes, block and stream ciphers (including AES) have a common denominator. This common feature is the exploitation (by various methods) of the existence of various types of algebraic relations that involve both the inputs and the outputs of some component. We will formulate the resulting design criteria on the respective components that will be very similar, if not identical.

## 2     From Boolean Functions to Algebraic Relations

Most of the current cipher design paradigms can be seen in terms of looking for in some sense "good" Boolean functions / "good" vectorial functions (S-boxes) and avoiding "bad" ones. The outputs of cryptosystems (and their components) should simply not depend on their inputs in a way that is too simple. The definition of the word "simple" does naturally vary from one place to another. For example in the design of stream ciphers, there are many so called "non-linearity" criteria, dictated by some (not always really practical) attacks. Building ciphers with such components allows to make sure that many (from real to very theoretical) attacks will not work very well on these ciphers. For example, in [25] Golic explains the criteria on the Boolean functions that should be used in stream ciphers. Obviously these criteria, to some extent being necessary in the design of good ciphers, are by far insufficient and nothing guarantees that a cipher that made out of "good" components will be good itself (i.e. will be secure). Moreover, using such components is sometimes even perceived (if they are really very good) as a potential danger (special may mean dangerous). In particular, many recent attacks in different areas of cryptography do work in spite of using very good (sometimes optimal) components w.r.t. aforementioned criteria (for example highly non-linear components).

### 2.1     Interesting Special Case: AES S-Box

AES (Rijndael) [19, 20] is precisely a good case to study in this respect. First, because its security is simply essential, and more importantly, because it pushes the (aforementioned) philosophy that culminates two decades of research in the design of modern ciphers to its limits. A general question is, whether it is possible (and how) to attack ciphers build with highly-nonlinear components (and thus build with eminently "good" Boolean function. Obviously studying this question will in most cases not give results being directly applicable to AES, but it gives us the opportunity to come up with new approaches to attack AES later, as well as should help us to simply design much better ciphers in the future (that avoid also the recent attacks).

In [5], Canteaut and Videau study the non-linearity properties of the Inverse function in $GF(2^n)$ (the only non-linear component of AES) with relation to linear, differential and higher-order differential attacks. It is exceptional and close to optimality, see [5]. On page 6 of [21], the designers of AES say: *"[...] The disadvantage of these boxes is that they have a simple description in $GF(2^m)$, [...] we are not aware of any vulnerability caused by this property. [...] Should such a vulnerability exist, one can always replace the Sboxes by Sboxes [...] that are not algebraic over $GF(2^m)$. [...]"* Unfortunately important vulnerability of the inverse S-box does exist. Historically the idea goes back to the algebraic attacks on several so called multivariate public key schemes, initiated by Patarin in [41], greatly improved by Courtois et al. [9, 18], and recently adopted by Faugère and Joux [31]. The seminal idea (due to Patarin) is to study the security of a cipher component not in terms of Boolean/algebraic functions, but in terms of Boolean/algebraic **relations** that involve both inputs and output bits. In the last two years, this precise idea, has led to a sudden collapse of several important families of stream ciphers, as demonstrated by Courtois, Meier et al in [16, 17, 2, 10, 12] and numerous other recent papers. We explain these in Section 4. But does it matter at all for block ciphers? This will the main subject of this paper starting from Section 5.

## 3    From Multivariate Public Key Schemes to General Algebraic Attacks

At Crypto'95, Jacques Patarin proposes a very interesting attack on the Matsumoto-Imai public-key cryptosystem of Eurocrypt'98, see [36, 40]. This cryptosystem, at the time considered as very promising, is based on a univariate transformation, that can be for example $X \mapsto X^3$. This cube function, instead of being over a ring of numbers modulo some $N$ like with RSA, is over a finite field, for example $GF(2^{80})$. The order of a multiplicative group of $GF(2^{80})$ is known and therefore in many cases, such a power function over a finite field is, unlike in RSA, easily invertible. However, the same algebraic structure of this function can be "concealed" (cf. [36, 40]) when it is written in a new representation, as a set of multivariate quadratic polynomial functions. It is done in such a way that it is easy to compute it forwards, and hard backwards, for anyone that does not known how the system of equations have been generated. Thus, Matsumoto and Imai construct their public key cryptosystem, see [36] for more details.

Incidentally, due to the cube function, this cryptosystem have extremely good properties when considered in terms of Boolean functions, see [39, 5]. Yet, this did not prevent Jacques Patarin from rather badly breaking this cryptosystem, at Crypto'95 [40]. He shows that there are simple algebraic relations that relate input and output bits of this cryptosystem. More precisely, if the input is $(x_0, \ldots, x_{79})$ and the output is $(y_0, \ldots, y_{79})$ there exist bi-linear equations of type, for example $\sum_{ij} \alpha_{ij} x_i y_j = 0$. Then, Patarin remarks that if such equations exist, they can be easily found from the public key, and then subsequently they can be used to decrypt any message: if we substitute a concrete values of $y$ in these equations they become linear and can be solved to recover the $x_i$.

This attack has been generalised by Courtois in [9]. This paper also proposes a first "theory" of algebraic attacks on public key schemes[1] that we will develop and explain here. This "theory" is quite simple and can potentially be applied to many different situations that arise in cryptanalysis. To achieve this we will be voluntarily imprecise. Some details vary from one attack to another, and it should be applicable also to situations that are very different than the area of algebraic attacks.

From one point of view, one can think that it applies to more or less all cryptographic attacks. To explain this, let's consider any attack on any deterministic one-way function which is described as a set of explicit arithmetic formulae $y_i = F_i(x_0, \ldots, x_{n-1})$. The answer $x$ we are looking for is also seen as a set of equations, though much simpler $x_i = \ldots$, which a hypothetical attack would evaluate to. We wish to look at any deterministic attack as a series of transformations that starts from (somewhat "complex") initial equations and eventually produces somewhat "simpler" ones (containing the solution to the system). Similarly, following [9], starting from some notion of complexity that is adapted to our initial equations, and makes them hard to solve, we can also try to construct attacks that work exactly in this way. For this, still following [9], we need to study (and find) methods that given some initial equations, give hope to generate some "simpler" equations. With such methods we hope to solve the system, by successive simplifications. For example, one possible notion of complexity is the non-linear degree. In Matsumoto-Imai and HFE systems [36, 40, 41] we have initial equations that are quadratic and our goal will be to find some simpler, linear equations. Most attacks known on these systems work in this way, e.g. [40, 41, 9, 18, 31].

Attacks that work in such a way, may be iterative with many steps, which makes them hard to understand and study. For example it is far from being clear what is the complexity of Courtois-Pieprzyk XSL attack on AES [15]. However, again following [9], what one should study, and what is really interesting, is what happens in one step of the attack. From the cryptological point of view the main question will be not what is the exact complexity of an attack, but rather if the attack is feasible in general (at least in some cases), and even more importantly, how to completely avoid such attacks. For these questions, the most important answers may already be given by looking at the beginning of the attack process. Do we gain something ? Can we by some means gain anything simpler from the initial equations ? Obviously it is always possible to combine equations in some way, (and it is very simple for Boolean algebraic equations over a finite filed). However, usually, we obtain other equations that have nothing special and are in fact more complex than the initial equations. Following [9], the interesting phenomenon to watch for is a type of "collapse in the complexity". For example, we take some multivariate equations of degree 2, combine them algebraically to get an equation of expected degree 4, but when we

---

[1] It applies also almost literally to algebraic attacks on block and stream ciphers, but at the time, nobody really suspected this.

compute this equation its degree collapses from 4 to 3. Here we gain something, some simplification arises. The heuristic is then that, if it can be done once, it can be done several times and in many cases we end up by obtaining a full working algebraic attack. In rare cases, it will obviously fail, but we know that designing systems such that there is no "collapse of complexity" in some sense, will prevent many attacks, whether they work well, or not. For example, building a cipher with large random components(e.g. S-boxes) makes such cases of "complexity collapse" to some degree very unlikely if not impossible, this whatever is our definition of complexity.

When, as in many cases studied in this paper, the notion of complexity is the non-linear degree of a multivariate polynomial form of a function, the existence of "complexity collapse" can be characterised as follows. If an algebraic combination of the original equations is of lower degree than expected, it means that there exist a non-trivial and in some sense "simple" (e.g. low degree) function $G$ such that:

$$G\left(x_0, \ldots, x_{n-1}; \quad F_0(x_0, \ldots, x_{n-1}), \ldots F_{m-1}(x_0, \ldots, x_{n-1})\right) = 0$$

If we replace $y_i = F_i(x_0, \ldots, x_{n-1})$ we get an algebraic relation between input and output bits:

$$G\left(x_0, \ldots, x_{n-1}; \quad y_0, \ldots, y_{m-1}\right) = 0 \quad (*)$$

In these formulas the $x_i$ and the $y_i$ may be in $GF(2)$, but may be also in any other finite field $GF(q)$. We are at the right point. It turns out that talking about algebraic relations is more general than considering "a collapse in the complexity": algebraic relations may exist, be found and directly be used in an attack, disregarding the initial complexity of the equations, that in some cases is within no comparison (much more complex).

Undoubtedly, there are many cases in which the very existence of an algebraic "complexity collapse" or/and resulting algebraic relations at some level, is somewhat trivial and inevitable. There are also many cases in which such occurrence can be an isolated phenomenon that does not lead to interesting attacks. Yet, to make sure that a system resists to large class of possible attacks it is sensible to avoid such situations whatsoever. (This concerns, as we will explain later mainly Generalised Linear Cryptanalysis and direct algebraic XSL-type attacks, and potentially other future attacks). Another way of seeing such design criteria is to say that, in a sense, components of our system (or the whole system itself) will be "more" indistinguishable from random components (e.g. random functions or random permutations), and thus less attacks should be possible.

In the following sections we will explain briefly, how this general paradigm of algebraic attacks applies to other contexts. This list is not exhaustive, and we expect that many other areas of cryptographic security can be described in a meaningful way in terms of "complexity collapse" and/or simple "I/O relations" with respect to some (not necessarily algebraic/polynomial degree) notion of complexity.

## 3.1    How to Build Secure Multivariate Public Key Cryptosystems

Here the conclusion follows immediately: for a trapdoor function to be secure we need to make sure that there is no multivariate relations such as (∗) that contain less than, let's say $2^{80}$ different monomials (in general, for finite systems, it is impossible to avoid the existence of algebraic relations, but their size will be astronomical). In practice, for most systems, if there is no algebraic/multivariate relations of size less than $2^{40}$, there should be no practical algebraic attack on the system (because we need to be able to recover the equations first). However, in some special cases, equations of large sizes can be build directly by a method that depends on the cipher, and then they can be used by substitution of variables. Therefore the proposed bound of $2^{80}$ gives a better guarantee.

# 4    Algebraic Attacks on Stream Ciphers

The algebraic attacks on stream ciphers have been introduced in 2003 by Courtois and Meier [17, 16]. Since then, the area has known an important research activity with many interesting contributions, to quote only some, by Armknecht and Krause [2, 1], Cho and Pieprzyk [6], Courtois [12, 10], Hawkes and Rose [27], Lee, Kim, Hong, Han and Moon [33], Meier, Carlet and Pasalic [37], and others. In this paper we only explain the main principle of algebraic attacks on stream ciphers from [17, 10], and what are the resulting design criteria for components of such ciphers.

The algebraic attack on stream ciphers is extremely general and applies potentially to all ciphers that have some linear feedback (for example based on LFSRs or cellular automata). We assume that in our cipher the first (linear) component is as follows. Let $x = (x_0, \ldots, x_{n-1}) \in GF(q)^n$ be the state of this component. We assume that the cipher is regularly clocked (some relaxations are possible, see [17, 16]) and at each clock the linear state $x$ is updated by some multivariate linear function $L$. This means that at each clock $x$ becomes $L(x)$, and if $K = (K_0, \ldots, K_{n-1})$ is the initial state, at time $t$ the state will be called $x^{(t)}$ and by definition we have $x^{(t)} = L^t(K)$.

Then we assume that the state of the linear component is supplied to the second "filter/combiner" component that outputs the keystream (it may output one or several bits at a time). This output component can be stateless or stateful: in the second case it also has internal memory bits that are updated at each clock. In this case, we have in addition to the linear feedback in the first component, a non-linear feedback in the second component (but usually of much smaller size/importance than the linear feedback).

Let $l$ be the number of memory bits in the second component, that after the time $t$ are $a_0^{(t)}, \ldots, a_{l-1}^{(t)}$. In particular, for stateless filters/combiners $l$ is 0, for example when a Boolean function is used to filter/combine the state bits of one or several LFSRs. The initial inner state is $a^{(-1)}$, exists before $t = 0$, and can be anything (it is unknown in the attack and algebraic attacks tend to eliminate all the monomials in the $a_i$). At each clock $t$, the combiner outputs $m$

bits $y_0^{(t)}, \ldots, y_{m-1}^{(t)}$, for $t = 0, 1, 2, \ldots$. For example, if the ciphers uses a single Boolean function to combine input bits, we have simply $m = 1$. In general, the second component can be described as a pair of functions $F = (F_1, F_2) : GF(2)^{n+l} \rightarrow GF(2)^{m+l}$, that given the current state and the input, computes the next state and the output:

$$
F : \begin{cases} (y_0^{(t+1)}, \ldots, y_{m-1}^{(t+1)}) = F_1(x_0^{(t)}, \ldots, x_{n-1}^{(t)}, a_0^{(t)}, \ldots, a_{l-1}^{(t)}) \\ (a_0^{(t+1)}, \ldots, a_{l-1}^{(t+1)}) = F_2(x_0^{(t)}, \ldots, x_{n-1}^{(t)}, a_0^{(t)}, \ldots, a_{l-1}^{(t)}) \end{cases}
$$

The most general form of an algebraic attack on stream ciphers following closely [10, 12, 17] works as follows.

- We assume that $L$ is known (for example the LFSRs used in the cipher are known or can be guessed/revovered).
- We consider $M$ consecutive states of the cipher.
- Find (by some method that is very different for each cipher) one (at least, but one is enough) multivariate relation $G$ between the state bits $x_i$ and some $M$ consecutive outputs, for example:

$$
G(x_0, x_1, \ldots, x_{n-1}; y^{(0)}, \ldots, y^{(M-1)}) = 0
$$

We assume that $G$ is of degree $d$ in the $x_i$ (the degree in the $y_i$ may also be important, but usually will not influence the total attack complexity).

- By recursive structure of the cipher, for any initial state $K$ and for any $t$, the same equation will apply to all consecutive windows of $M$ states

$$
G(L^t(K); y^{(t)}, \ldots, y^{(t+M-1)}) = 0
$$

- The $y^{(t)}, \ldots, y^{(t+M-1)}$ are replaced by their values known from the observed output of the cipher.
- Due to the linearity of $L$, for any $t$, the degree of these equations is still $d$.
- For each keystream bit, we get a multivariate equation of degree $k$ in the $x_i$.
- Given many keystream bits, we inevitably obtain a very overdefined system of equations (i.e. great many multivariate equations of degree $d$ in the $K_i$).
- To solve these equations we may apply the XL algorithm from Eurocrypt 2000 [43], adapted for this purpose in [16] and other improved elimination techniques such as computing Gröbner bases combined with linear algebra, see [22, 23]. However, if we dispose of a sufficient amount of keystream, (which is frequently not very big, see [17]), all these are not necessary.
- If the amount of keystream available is large enough, we use a so called linearization method that is particularly simple. There are about $T \approx \binom{n}{d}$ monomials of degree $\leq d$ in the $n$ variables $K_i$ (assuming $d \leq n/2$). We consider each of these monomials as a new variable $V_j$. Given about $\binom{n}{d} + M$ keystream bits, and therefore $R = \binom{n}{d}$ equations on successive windows of $M$ bits, we get a system of $R \geq T$ linear equations with $T = \binom{n}{d}$ variables $V_i$ that can be easily solved by Gaussian elimination on a linear system of size $T$. The time to solve such a system is $T^\omega$ with in theory $\omega \leq 2.376$ [7] but in practice for small systems, it is believed that one should rather consider $\omega$ that is closer to 3 than to 2.376.

## 4.1   How to Build Secure Stream Ciphers

For stream ciphers in which the second component does not have internal memory, the case $M > 1$ does not make a lot of sense, and if we wish the cipher to avoid algebraic attacks, we get a requirement on the second component that is identical to our requirement on public key trapdoor functions. There should be no "simple" algebraic relations between its inputs and outputs such as:

$$G(x_0, \ldots, x_{k-1}; \; y_0, \ldots, y_{m-1}) = 0 \quad (*)$$

Similarly, in the general case $l \geq 1$ we need to avoid the existence of "not too complex" equations (that eliminate the internal state bits $a_i$) of type:

$$G(x_0, x_1, \ldots, x_{n-1}, \; y^{(0)}, \ldots, y^{(M-1)}) = 0 \quad (**)$$

For stream ciphers however, the notion of "simple" and "complex" equations changes. It is no longer the total size of these equations (number of monomials) that matters, but their degree in the $x_i$ (their degree in the $y_i$ can be large, provided that the total size of the equations is not too big and that there is some method to generate these equations from the description of the cipher). Our recommandation, for ciphers that aim at $2^{128}$ security is that there should be no $G$ that can be **efficiently written** (for example using up to $2^{128}$ of memory) with degree $d \leq 16$. (We do not exactly require that they do not exist, and for some high $d$ there may exist large relations with, for example $2^{100}$ monomials, that are not a problem as long as there is no efficient algorithm to recover/write and otherwise use them). For higher security levels, for example military-level requirements of type $2^{256}$, we recommend a cautious $d \geq 32$. For specific ciphers these numbers may be lower but then they require a careful study if they will not be broken by fast algebraic attacks [12, 1, 27].

It is certainly possible to obtain components that satisfy these criteria by using sufficiently large random S-boxes (the exact size will depend a lot of the exact construction). Otherwise, proposing constructive methods to obtain components that will (if possible provably) satisfy these criteria is an important open problem. For Boolean functions, this problem can be rephrased as constructing "good" Boolean function that in addition to classical non-linearity criteria respond also to the new criterion of "algebraic immunity". It also remains an open problem, see [37].

## 5   Block Ciphers and Algebraic Relations

This paper is about a simple idea of studying algebraic relations on different components. In this paper we will not try to summarise all the results but the outcome of this approach on stream ciphers and multivariate public key schemes was quite devastating, see among others [1, 2, 6, 9, 10, 11, 12, 14, 15, 16, 17, 18, 22, 23, 27, 31, 33, 37, 40, 41, 42, 43]. Several classes of schemes were shown to be substantially less secure than expected, and sometimes badly broken. But the real question that many people are asking is, does this type of attacks matter also for block ciphers ?

At present many cryptologists still believe that they don't matter (at all). Yet, from one point of view there is no doubt that it does ! For example with the polynomial approximation attack of [30], Jakobsen was the first to claim that to obtain secure ciphers *"[...] it is not enough that round functions have high Boolean complexity. [...]"* . He proposes already to avoid functions that have simple algebraic properties in the design of block ciphers (but his warning was never taken seriously). Regarding the AES S-box, in [8] and in [13] in these proceedings, Courtois shows that it is possible to construct, by several very different methods, many block ciphers based on the inverse in $GF(2^n)$ that satisfy all the known design criteria on block ciphers, yet remain very very weak.

These schemes are insecure, because the Inverse-based Rijndael-type S-boxes, though very complex when regarded as a function, can be characterised in several ways by algebraic relations, cf. [13, 15, 38]. Here we are concerned with attacks being forms of generalised linear cryptanalysis, see [26, 32, 8, 13]. Though these attacks techniques clearly do evolve into general attacks that can be applied potentially to any block cipher, the insecure ciphers constructed in [13] remain very special contrived ciphers.

On the contrary, for ciphers such as DES and AES, that use relatively small S-boxes and a lot of diffusion that connects the outputs of one S-box to many other S-boxes in the next rounds, (wide trail strategy of AES designers [19, 20]), we expect that the things should be very different. In [13, 8], heuristic arguments are given to the effect that, the impact of generalised linear cryptanalysis on such ciphers (e.g. AES) is expected to be low, as long as they resist well to linear cryptanalysis. Therefore, it seems so far that the algebraic relations may do not really matter so much for AES and similar ciphers.

# 6    Global Algebraic Attacks on Block Ciphers

We see that, finding attacks on ciphers such as AES, remains an ambitious task, even given the existence of algebraic relations on the S-boxes. Unfortunately, there is yet another attack strategy, published in 2002 by Courtois and Pieprzyk, that is designed to render the "wide trail strategy" useless. It can be called a **direct/global** algebraic attack strategy, or **exact** algebraic approach. At the origin, it also uses the existence of algebraic relations for the individual components of the cipher. We do not however try to connect the specific monomials that appear in one equation to another equation, which may be very hard, but just write the equations for the whole cipher, to obtain a global system of equations that uniquely characterizes the key to be found. Then we see if it is possible (in theory and/or in practice) to solve such a system of equations.

This type of approach, if proven to work efficiently in practice, is not less than a major revolution in the field of block cipher cryptanalysis. This is because, except few very weak ciphers, all the general attacks known up till now for block ciphers are attacks that combine "approximations", that are some properties (linear, differential, higher-order differential, polynomial approximation etc..) true with some probability that except for some very weak ciphers is different from 1. This "combine approximations" paradigm has three important consequences. First of all, the complexity of the attacks must grow exponentially with the number of rounds. Secondly, the number of plaintexts needed in an attacks also grows in the same way (and may be the main limitation in practice). Finally, ciphers with good diffusion (wide trail strategy) force the attacker to use several approximations in parallel in the same round, and the efficiency of the attacks further decreases.

The "exact algebraic" approach that exploits equations that are true with probability 1 that exist locally (for example for each S-box) has the potential to remove simultaneously the three aforementioned obstacles. The complexity is not longer condemned to grow exponentially with the number of rounds. The number of required plaintexts may be quite small (e.g. 1). And the wide trail strategy should have no impact whatsoever on the complexity of the attack.

## 6.1    How Secure Are Today's Block Ciphers ?

Some people dismiss the idea of an algebraic attack on AES, as being too simple and too naive to be true. Our impression is that, it is rather the current thinking about the security of block ciphers that is very naive.

We get the impression that, if we mix sufficiently many rounds of any construction, it will be secure. In practice however, the ciphers are meant to be rather fast, have a limited number of rounds, but yet the security claims made on them are extremely ambitious. During the AES contest many authors proposed ciphers claimed to be indistinguishable from a random permutation within less than $2^{256}$ computations. This is a huge number. With the Moore's law, such keys should remain secure against brute force until around 2200. This gives us 200 years to invent new mathematics, new algorithms, and new attacks that will break the cipher faster than the exhaustive search before it is outdated. Who can make security predictions for such a long period of time, knowing that so many security claims are disproved every year ? Moreover, $2^{256}$ is close to the number of atoms in the universe, therefore it also possible that the computers will never actually have such a computing power. This means that we are left with infinite time to find better attacks. We believe therefore that betting that a cipher cannot be distinguished from a random cipher faster than by brute force, may be an infinitely risky bet for 256-bit ciphers. Our guess is rather that **all** the block ciphers with 256-bit keys that were submitted to AES, will some day be broken faster than by exhaustive search, simply because our current knowledge about the real security of block ciphers is yet very low.

## 6.2    Who Invented Algebraic Attacks on Block Ciphers ?

According to a visionary recommandation of Shannon from his 1949 paper [44], breaking a good cipher should require: *"as much work as solving a system of simultaneous equations in a large number of unknowns of a complex type".* There are many ways of interpreting this statement. For example we may think about multivariate quadratic equations with Boolean variables, the large number of unknowns may mean a large number of monomials, unknowns of a complex type may mean monomials of high degree (or that combine variables that come from remote location inside a cipher).

From another point of view, it is a trivial folklore attack that anyone can think of. Indeed, it is easy to see that, for any practical cryptographic system that relies on computational (not information-theoretic) security, we can write a system of Boolean equations such that solving it allows to find the key. Then, solving a general system of Boolean equations is an NP-hard problem, and solving non-linear systems of large size is expected to be hopeless. However, it turns out that, what makes such problems hard is not so much the number of variables or monomials, but the balance between the number of equations and the number of monomials that appear in these equations. From this, we expect that, systems that are overdefined, sparse, or both, should be much easier to solve than general systems of similar size. As far as we know, before 1998-2000, the scientific community were not aware of this fact, and easily believed that large systems of equations are necessarily hard to solve. When the XL attack was first introduced by Courtois, Klimov, Patarin and Shamir [43], as a development of earlier ideas of Shamir and Kipnis [42], things started to change. In particular, specialists of elimination methods such as Gröbner bases that have been studied for many years now, see for example [45, 22, 23], started to realise the full potential of these and other algebraic techniques to solve problems that arise in cryptography. It turns out that the cryptographic instances of multivariate systems of equations have several interesting properties that may and do help to solve them efficiently. Among these properties we will quote the fact that they are over very small finite fields, they usually have a unique solution, they do not have solutions in extensions fields or at infinity, and again, they are frequently over-determined, and sparse (with several possible notions of sparsity).

At present the area of algebraic attacks is full of open problems that should be solved with time. A lot remains to be done in discovering cryptanalytic applications of already existing algebraic methods of solving systems of polynomial equations. Similarly, specific systems of equations that arise in cryptography should allow (and already do) to better understand why certain very general algebraic algorithms (such as Buchberger or F5 algorithms) for solving equations work well in some cases, and do fail in some other cases. Finally, new methods of solving algebraic equations should and will be invented, motivated by cryptographic attacks.

## 6.3   The Structure of Algebraic Attacks

Global algebraic attacks on block cipher following Courtois and Pieprzyk (previously imagined also by Shannon, Patarin and probably few others) contains the following three stages, that can and should be studied separately.

1. **Write an appropriate initial system.** Write a system of equations that, given one or several known plaintexts, uniquely characterizes the key. This

system should be as over-determined (also called overdefined) and as sparse as possible. This can be measured by the initial ratio $R_{ini}/T_{ini}$ between the number of equations $R_{ini}$ in the system and the total number of monomials $T_{ini}$ that appear in it. It can be for example 1/4 or 1/3. It is not clear what is the optimal setting for algebraic attacks: we may try simply to achieve a lowest $R_{ini}/T_{ini}$ possible, however for some systems with a higher initial ratio, but a lower global size, or some specific additional properties, the overall complexity of an algebraic attack may be lower.

2. **Expand it.** The second step is an expansion step. The goal is, starting from the original $R_{ini}$ equations with $T_{ini}$ monomials, to produce (for example by multiplying the equations by some well chosen polynomials) another (much bigger) set of $R$ equations with $T$ monomials. The goal is to have the new ratio $R/T$ close (or bigger than) 1. If $R > T$ it means that the set of equations is redundant, and we should think of a better method of generating them (to avoid redundancies) and also of a better method of counting how many equations we have, that are not trivial linear combinations of other equations, and therefore serve no purpose.

Here the main criterion of "success" is not so much the final ratio $R/T$ (that simply must be somewhat close to 1, e.g. 0.9) but the size $T$. However it remains possible that some attacks with a worse (larger) $T$ and better (bigger) $R/T$ do in fact work better (cf. next step).

3. **Final in place elimination.** The final step should be an "in place" elimination method that given an "almost saturated system" with $R/T$ close to 1, finds a solution. On proposed method to achieve this is by generating a completely saturated system (the T' method proposed by Courtois in [15, 14]. It can also be achieved by computing a Gröbner basis of the expanded system, and probably by other means. The (heuristic) requirement is that the memory required in this third step should not exceed $T$, otherwise maybe we need to improve rather the second (expansion) step.

## 6.4    Applicability of Algebraic Attacks

There are reasons why, overdefined and/or sparse systems are bound to appear frequently in cryptography. In most settings, there is no cryptographic solutions with unconditional security, and we have to rely on computational security. A relatively short (128 bits or less) key will be usually used many times, to produce much more information: many known plaintexts, many signatures, etc. In public key cryptography, a proof of security would allow to be certain that each utilization of the cryptographic scheme, does not leak useful information. Secret key schemes do not have such proofs of security, and the more we use it, the more the problem become overdefined (if we do not introduce additional variables). It is also in secret key cryptography, that the problems may become really massively overdefined, if we think about the amounts of data that can be encrypted with a single key, on a satellite link. Another problem is a consequence of the fact that many ciphers are designed to be implemented in hardware with a very low gate count. This allows to design an algebraic attack with relatively small umber of variables and a very small number of monomials (very sparse).

These are theoretical considerations. The present experience of algebraic attacks is that, their complexity should grow "nearly polynomially" in the number of rounds and in the block size, with however a really huge constant called $\Gamma$ that does depend only on the S-box. (This for all known versions of the XSL attack, and for both resulting definitions of $\Gamma$, see [15]). For a random S-box (and also for many other S-boxes that have no special properties such as algebraic relations) this constant $\Gamma$ can be shown to be double-exponential in $s$, the size of the S-box in bits. In [15], it appears that already 4-bit S-boxes, should be sufficient for $2^{128}$ security and probably beyond. For the Rijndael S-boxes, it is possible to see that $\Gamma$ grows only simply exponentially in $s$. Then it seems that even for $s = 8$ algebraic attacks faster than $2^{128}$ may exist, see [15, 38], but we are clearly on the frontier of applicability of algebraic attacks. Thus, it seems that in fact algebraic attacks are only possible for some very special ciphers. Apart from Serpent and Rijndael, we are not aware of a single other block cipher for which even a current (probably too optimistic) estimations of the complexity of algebraic attacks would give less than the exhaustive search.

## 6.5    Is AES Broken ?

It is important to say: we really do not know. It is possible that, one of the XSL attacks works exactly as predicted, or a simple combination of already known attacks already breaks AES. Our favorite candidate in this respect would be to combine the Murphy-Robshaw idea of using equations over $GF(256)$ from [38], with one of the XSL expansion attacks from [15], and replace the final $T'$ method by a (presumably better) advanced Gröbner bases algorithm such as Faugère's F5 [23]. This might simply break AES. But it is also possible that it fails quite miserably for some fundamental reason that is not yet understood. Then, a slight modification of the attack could still remove the theoretical obstacle and give an attack that might again work in practice. Studying algebraic attacks on block ciphers in all due details is outside the scope of this paper, and remains largely to be done. Both theoretical and experimental results will probably be needed to get the full picture.

## 6.6    How to Avoid Algebraic Attacks on Block Ciphers

At any rate, we advocate to take the algebraic attacks on block ciphers very seriously and to design block ciphers that do avoid such attacks. The resulting security criterion is, still more or less the same. The S-boxes of a block cipher should avoid the existence of "simple" algebraic relations of type:

$$G\left(x_0, \ldots, x_{s-1};\; y_0, \ldots, y_{s-1}\right) = 0 \quad (*)$$

The exact definition of "simple" that would prevent all algebraic attacks on block ciphers is not obvious to give. We need to avoid equations that, for some representation, and some system of equations, give a low value of $\Gamma$. For example following [15], we should avoid systems that are too overdefined or/and too sparse.

This should not be very hard to achieve. We believe that using random S-boxes on 8 bits should be about sufficient to achieve 128-bit security (though not for sure). We recommend in fact to construct bigger S-boxes that have no algebraic relations starting from random bijective 8-bit S-boxes. For higher security requirements such as military applications, we advocate to make mandatory a requirement that the cipher should use at several places inside the encryption, a random S-box of at least 16 bits.

## 7   The Future of AES

In our opinion, AES should still be recommended as the best choice of encryption algorithm for applications that do not require long-term security. We believe however that NIST should set an expiration date for AES, that could be 2010. It could be extended it later, according to the developments in cryptanalysis, but we believe that in 2010 it will be much wiser to replace AES by a better cipher, being not vulnerable to algebraic attacks, generalised linear cryptanalysis with multiple approximations, and other attacks that will probably be invented by 2010. The replacement should be done even if it turns out that known algebraic attacks on block ciphers do not work, and all other attacks that exploit algebraic relations (e.g. generalised linear cryptanalysis) do not break AES either.

In addition, we believe that a cipher such as AES can only be really credible as the world's standard all-purpose cryptographic high security lock, if there is a series of AES challenges. They could range from 100 to 1 million dollars, and be offered for solving various important open problems that in a different manner do compromise the security of AES, up to a total break that is done or doable in practice. This would allow to monitor the progress in the security of AES and to ascertain a serious status of this scheme compared to so many other schemes that are broken every year. For people that do not have expertise in cryptography, and cannot tell between real or fake security experts, such challenges, are the only way of knowing that the AES is indeed not yet broken, and some people take its security seriously enough to offer 1 million dollar to whoever demonstrate it can be broken in practice.

## 8   Conclusion

Algebraic attacks exploit the existence of multivariate relations on the appropriate cryptographic component. They do allow to break many multivariate public key schemes and stream ciphers. For block ciphers, their effectiveness is far from being clear. Yet, it is very sensible to avoid the existence of such algebraic relations for non-linear components of block ciphers. This not only because of algebraic attacks, but also because of generalised linear attacks: examples of contrived ciphers are known that are not secure with relation to these.

Thus, we propose (if possible) to simply avoid multivariate and algebraic relations in **all** types of cipher components. This extends the current paradigm of avoiding "bad" Boolean functions, or/and "bad" vectorial functions (S-boxes).

One method to achieve this would be to construct appropriate cryptographic components with guaranteed "algebraic immunity". A much simpler method, is to use sufficiently large random S-boxes. This should prevent all known attacks on block ciphers: linear/differential cryptanalysis with generalisations, all kinds of generalised linear attacks as described in [13], and also any kind of exact algebraic attacks such as XSL [15].

# References

1. Frederik Armknecht: *Improving Fast Algebraic Attacks*, to appear in FSE 2004, LNCS, Springer.
2. Frederik Armknecht, Matthias Krause: *Algebraic Atacks on Combiners with Memory*, Crypto 2003, LNCS 2729, pp. 162-176, Springer.
3. Kazuaro Aoki and Serge Vaudenay: *On the Use of GF-Inversion as a Cryptographic Primitive. SAC 2003, LNCS 3006, pp. 234-247, Springer 2004.*
4. *Ross Anderson, Eli Biham and Lars Knudsen: Serpent: A Proposal for the Advanced Encryption Standard.*
5. Anne Canteaut, Marion Videau: *Degree of composition of highly nonlinear functions and applications to higher order differential cryptanalysis*, Eurocrypt 2002, LNCS 2332, Springer.
6. Joo Yeon Cho and Josef Pieprzyk; *Algebraic Attacks on SOBER-t32 and SOBER-128*, will appear in FSE 2004, LNCS, Springer.
7. Don Coppersmith, Shmuel Winograd: *Matrix multiplication via arithmetic progressions*, J. Symbolic Computation (1990), 9, pp. 251-280.
8. Nicolas Courtois: *Feistel Schemes and Bi-Linear Cryptanalysis*, To be presented at Crypto 2004, Santa Barbara, California, 15-19 August 2004.
9. Nicolas Courtois: *The security of Hidden Field Equations (HFE)*; Cryptographers' Track Rsa Conference 2001, LNCS 2020, Springer, pp. 266-281.
10. Nicolas Courtois: *Algebraic Attacks on Combiners with Memory and Several Outputs*, Available on http://eprint.iacr.org/2003/125/. 23 June 2003.
11. Nicolas Courtois: *La sécurité des primitives cryptographiques basées sur les problèmes algébriques multivariables MQ, IP, MinRank, et HFE*, PhD thesis, Paris 6 University, September 2001, in French. Available at http://www.minrank.org/phd.pdf.
12. Nicolas Courtois: *Fast Algebraic Attacks on Stream Ciphers with Linear Feedback*, Crypto 2003, LNCS 2729, pp: 177-194, Springer.
13. Nicolas Courtois: *The Inverse S-box, Non-linear Polynomial Relations and Cryptanalysis of Block Ciphers*, in AES 4 Conference, Bonn May 10-12 2004, LNCS, Springer.
14. Nicolas Courtois and Jacques Patarin, *About the XL Algorithm over GF(2)*, Cryptographers' Track RSA 2003, LNCS 2612, pages 141-157, Springer 2003.
15. Nicolas Courtois and Josef Pieprzyk, *Cryptanalysis of Block Ciphers with Overdefined Systems of Equations*, Asiacrypt 2002, LNCS 2501, pp.267-287, Springer, a preprint with a different version of the attack is available at http://eprint.iacr.org/2002/044/.

16. Nicolas Courtois: *Higher Order Correlation Attacks, XL algorithm and Cryptanalysis of Toyocrypt*, ICISC 2002, LNCS 2587, pp. 182-199, Springer.

17. Nicolas Courtois and Willi Meier: *Algebraic Attacks on Stream Ciphers with Linear Feedback*, Eurocrypt 2003, Warsaw, Poland, LNCS 2656, pp. 345-359, Springer. An extended version is available at http://www.minrank.org/toyolili.pdf

18. Nicolas Courtois, Magnus Daum and Patrick Felke: *On the Security of HFE, HFEv- and Quartz*, PKC 2003, LNCS 2567, Springer, pp. 337-350. The extended version can be found at http://eprint.iacr.org/2002/138/.

19. Joan Daemen, Vincent Rijmen: *AES proposal: Rijndael*, http://csrc.nist.gov/encryption/aes/rijndael/Rijndael.pdf

20. Joan Daemen, Vincent Rijmen: *The Design of Rijndael. AES - The Advanced Encryption Standard*, Springer-Verlag, Berlin 2002. ISBN 3-540-42580-2.

21. Joan Daemen, Vincent Rijmen, Bart Preneel, Anton Bosselaers, Erik De Win: *The Cipher SHARK*, FSE 1996, Springer.

22. Jean-Charles Faugère: *A new efficient algorithm for computing Gröbner bases ($F_4$)*, Journal of Pure and Applied Algebra 139 (1999) pp. 61-88. See www.elsevier.com/locate/jpaa

23. Jean-Charles Faugère: *A new efficient algorithm for computing Gröbner bases without reduction to zero (F5)*, Workshop on Applications of Commutative Algebra, Catania, Italy, 3-6 April 2002, ACM Press.

24. Niels Ferguson, Richard Schroeppel and Doug Whiting: *A simple algebraic representation of Rijndael*, SAC 2001, page 103, LNCS 2259, Springer.

25. Jovan Dj. Golic: *On the Security of Nonlinear Filter Generators*, FSE'96, LNCS 1039, Springer, pp. 173-188.

26. C. Harpes, G. Kramer, and J. Massey: *A Generalization of Linear Cryptanalysis and the Applicability of Matsui's Piling-up Lemma*, Eurocrypt'95, LNCS 921, Springer, pp. 24-38. http://www.isi.ee.ethz.ch/ harpes/GLClong.ps

27. Philip Hawkes, Gregory Rose: *Rewriting Variables: the Complexity of Fast Algebraic Attacks on Stream Ciphers*, by Philip Hawkes and Gregory G. Rose. In crypto 2004, to appear in LNCS, Springer, 2004. Available from eprint.iacr.org/2004/081/.

28. Thomas Jakobsen and Lars Knudsen: *Attacks on Block Ciphers of Low Algebraic Degree*, Journal of Cryptology 14(3): 197-210 (2001).

29. Thomas Jakobsen: *Higher-Order Cryptanalysis of Block Ciphers*. Ph.D. thesis, Dept. of Math., Technical University of Denmark, 1999.

30. Thomas Jakobsen: *Cryptanalysis of Block Ciphers with Probabilistic Non-Linear Relations of Low Degree*, Crypto 98, LNCS 1462, Springer, pp. 212-222, 1998.

31. Antoine Joux, Jean-Charles Faugère: *Algebraic Cryptanalysis of Hidden Field Equation (HFE) Cryptosystems Using Gröbner Bases*, Crypto 2003, LNCS 2729, pp. 44-60, Springer, 2003.

32. Lars R. Knudsen, Matthew J. B. Robshaw: *Non-Linear Characteristics in Linear Cryptoanalysis*. Eurocrypt'96, LNCS 1070, Springer, pp. 224-236, 1996.

33. Dong Hoon Lee, Jaeheon Kim, Jin Hong, Jae Woo Han and Dukjae Moon: *Algebraic Attacks on Summation Generators*, on eprint.iacr.org/2003/229/ and to appear in FSE 2004, LNCS, Springer.

34. R. Lidl, H. Niederreiter: *Finite Fields*, Encyclopedia of Mathematics and its applications, Volume 20, Cambridge University Press.

35. Alfred J. Menezes, Paul C. van Oorschot, Scott A. Vanstone: *Handbook of Applied Cryptography*; CRC Press, 1996.

36. Tsutomu Matsumoto, Hideki Imai: *Public Quadratic Polynomial-tuples for efficient signature-verification and message-encryption*, Eurocrypt'88, Springer 1998, pp. 419-453.
37. Will Meier, Enes Pasalic and Claude Carlet: *Algebraic Attacks and Decomposition of Boolean Functions,* Eurocrypt 2004, pp. 474-491, LNCS 3027, Springer, 2004.
38. S. Murphy, M. Robshaw: *Essential Algebraic Structure within the AES,* Crypto 2002, Springer.
39. Kaisa Nyberg: *Differentially Uniform Mappings for Cryptography,* Eurocrypt'93, LNCS 765, Springer, pp. 55-64.
40. Jacques Patarin: *Cryptanalysis of the Matsumoto and Imai Public Key Scheme of Eurocrypt'88*; Crypto'95, Springer, LNCS 963, pp. 248-261, 1995.
41. Jacques Patarin: *Hidden Fields Equations (HFE) and Isomorphisms of Polynomials (IP): two new families of Asymm. Algorithms,* Eurocrypt'96, Springer, pp. 33-48.
42. Adi Shamir, Aviad Kipnis: *Cryptanalysis of the HFE Public Key Cryptosystem*; In Advances in Cryptology, Proceedings of Crypto'99, Springer, LNCS. The paper can be found at `http://www.hfe.info`.
43. Adi Shamir, Jacques Patarin, Nicolas Courtois, Alexander Klimov, *Efficient Algorithms for solving Overdefined Systems of Multivariate Polynomial Equations,* Eurocrypt'2000, LNCS 1807, Springer, pp. 392-407.
44. Claude Elwood Shannon: *Communication theory of secrecy systems,* Bell System Technical Journal 28 (1949), see in patricular page 704.
45. Wang, D. *Elimination Methods,* Texts and Monographs in Symbolic Computation, Springer, 2001. XIII, ISBN 3-211-83241-6.

# An Algebraic Interpretation of $\mathcal{AES}-128$

Ilia Toli and Alberto Zanoni

Dipartimento di Matematica *Leonida Tonelli*,
Università di Pisa,
Via Buonarroti 2, 56127 Pisa, Italy
{toli, zanoni}@posso.dm.unipi.it

**Abstract.** We analyze an algebraic representation of $\mathcal{AES}-128$ as an embedding in $\mathcal{BES}$, due to Murphy and Robshaw. We present two systems of equations $S^*$ and $K^*$ concerning encryption and key generation processes. After some simple but rather cumbersome substitutions, we should obtain two new systems $\mathcal{C}_1$ and $\mathcal{C}_2$. $\mathcal{C}_1$ has 16 very dense equations of degree up to 255 in each of its 16 variables. With a single pair $(p, c)$, with $p$ a cleartext and $c$ its encryption, its roots give all possible keys that should encrypt $p$ to $c$. $\mathcal{C}_2$ may be defined using 11 or more pairs $(p, c)$, and has 16 times as many equations in 176 variables. $K^*$ and most of $S^*$ is invariant for all key choices.

## 1 Introduction

The well famous symmetric-key 64-bit cryptosystem $\mathcal{DES}$ [10] was broken in 1998 by means of a special purpose computer called $\mathcal{DES}$ *Cracker*. This computer contained 1536 chips, could search 88 billion keys/sec, and costed 250.000 \$. It won $\mathcal{RSA}$ Laboratories $\mathcal{DES}$ *Challenge II-2* by successfully finding a $\mathcal{DES}$ key in 56 hours in July 1998. In January 1999, $\mathcal{RSA}$ Laboratories $\mathcal{DES}$ *Challenge III* was solved by the $\mathcal{DES}$ Cracker working in conjunction with a worldwide network of 100.000 computers known as distributed.net. This cooperative effort found a $\mathcal{DES}$ key in 1335 minutes, testing over 245 billion keys/sec.

Other than exhaustive key search, the two most important attacks to $\mathcal{DES}$ are differential cryptanalysis and linear cryptanalysis. An actual implementation of the latter was carried out in 1994 by its inventor, Matsui. It is a known-plaintext attack using $2^{43}$ plaintext-ciphertext pairs, all of which are encrypted using the same, unknown, key. It took 40 days to generate them, and it took 10 days to actually find the key.

This cryptanalysis did not have any practical impact on the security of $\mathcal{DES}$, however, due to the extremely huge number of plaintext-ciphertext pairs that are required to mount the attack. It is unlikely in practice that an adversary will be able to accumulate such a huge number of plaintext-ciphertext pairs that are all encrypted using the same key.

On January 2, 1997, NIST began the process of choosing a replacement for $\mathcal{DES}$. The replacement should be called $\mathcal{AES}$ (*Advanced Encryption Standard*).

H. Dobbertin, V. Rijmen, A. Sowa (Eds.): AES 2004, LNCS 3373, pp. 84–97, 2005.

After a 3-years-long evaluation, Rijndael [3,4] was chosen to be the $\mathcal{AES}$ among 15 eligible candidates upon 21 submissions. It was published as FIPS 197 [5] on November 26, 2001.

Rijndael is a block cipher, that encrypts blocks of 128, 192, and 256 bits using symmetric keys of 128, 192, and 256 bits. It was designed with a particular attention to bit-level attacks, such as *linear* and *differential cryptanalysis*. What makes it particularly resistant to such attacks is the tension between operations in the two fields $\mathbb{F} = GF(2^8)$ and $GF(2)$. Since its proposal, several new bit-level attacks, such as *impossible differential* and *truncated differential* have been proposed. Most of them break with some efficiency reduced versions of Rijndael. For a general version of it, they are not much better than exhaustive key search. In practice they are mostly academic arguments rather than real world threats to the security of $\mathcal{AES}$. The interested reader can find an account and some references about these cryptological tools in [9].

Actually, only the blocklength 128 of Rijndael was approved to become $\mathcal{AES}$.

Another, new, cryptological tool is the algebraic representation of the cipher [8], [6], [2]. In this case, an eavesdropper tries to write the whole set of operations and parameters of the cipher by means of a system of polynomial equations, which he/she next tries to solve. In general, the related systems are enormous. Solving them by means of general purpose techniques, such as Gröbner bases [1] is considered the wrong way to face the problem. However, they sometimes bear a lot of intrinsic structure, that probably facilitates the task. A little research is done in the topic. Specially $\mathcal{AES}$ seems to have been designed regardless to algebraic cryptanalysis tools.

In this paper we focus on the $\mathcal{BES}$ algebraic approach, due to Murphy and Robshaw [8].

## 2    An Overview on $\mathcal{AES}$-*128*

Here is a sketch of $\mathcal{AES}$ encryption algorithm.

- Input a plaintext **x**. Initialize **State** = **x**, and perform an operation **AddRoundKey**, in w= hich we xor the **RoundKey** with the **State**.
- For each round but the last one, perform a substitution operation called **SubBytes** on **State**, using an $S$-box, perform a permutation **ShiftRows** on **State**, perform an operation **MixColumns** on **State**, and perform **AddRoundKey**.
- Perform **SubBytes**, **ShiftRows**, and **AddRoundKey**.
- Define the ciphertext **y** to be **State**.

All of $\mathcal{AES}$ operations are byte-oriented. The cleartext, ciphertext, and each output of mid-steps of encryption and decryption algorithms are thought of as $4 \times 4$ matrices of bytes. The arithmetic operations performed on each byte are those of the finite field $\mathbb{F} = GF(2^8)$. The elements are thought of as univariate polynomials with coefficients in $GF(2)$, $\mod (m(t))$, the so-called *Rijndael polynomial*:

$$m(t) = t^8 + t^4 + t^3 + t + 1 = \mathtt{11b} \tag{1}$$

**Fig. 1.** The **ShiftRows** operation on $\mathcal{AES}$

They are represented as couples of integers in hexadecimal representation. If interpreted as eight-bit binary strings, the numbers give the exponents of the terms in the polynomial representation.

The **SubBytes** operation substitutes each of the given bytes $x$ with $\mathcal{S}(x)$:

$$\mathcal{S}(x) = 63 + 8fx^{127} + b5x^{191} + 01x^{223} + f4x^{239} + 25x^{247} + f9x^{251} + 09x^{253} + 05x^{254} \tag{2}$$

Actually, $\mathcal{S}(x)$ is a permutation polynomial.

The **ShiftRows** operation permutes bytes in each row, as shown in Figure 1.

The **MixColumns** operation performs a permutation of bytes in each column by means of a certain matrix from the linear group $GL(\mathbb{F}, 16)$, explicitly given later in section 3. In practice, the columns are considered as polynomials from $\mathbb{F}[x]$, and multiplied $\mod(x^4 + 1)$ with the polynomial $a(x)$:

$$a(x) = 03x^3 + 01x^2 + 01x + 02. \tag{3}$$

### 2.1   The Key Schedule

The key used in every cipher round is successively obtained by the key of the precedent one. The whole procedure is sketched below.

- Input a key $\mathbf{h}_0$. Initialize $\mathbf{H}_0 = \mathbf{h}_0$.
- For each round $r = 1, \ldots, 10$, perform a permutation called **RotWord** on the sub-vector formed by the last four elements (word) of $\mathbf{H}_{r-1}$, as shown in Figure 2.
- Perform the **SubWord** ($\mathcal{S}$-box on each byte) operation on the obtained word, and add the constant vector $\mathbf{Rcon}_r = (t^{r-1}, 0, 0, 0)$. Define the other elements by means of bitwise **xor** operations in terms of the obtained result and other words from $\mathbf{H}_{r-1}$.
- Define the complete set of keys $\mathbf{h}$ to be $\{\mathbf{H}_r \mid r = 0, \ldots, 10\}$.

We consider each vector as divided into four words, indicated with a second index ranging from 0 to 3. For $\mathbf{y} \in \mathbb{F}^4$ we put

$$\varphi_A^r(\mathbf{y}) = \mathbf{SubWord}(\mathbf{RotWord}(\mathbf{y})) + \mathbf{Rcon}_r,$$

| $a_0$ | $a_1$ | $a_2$ | $a_3$ |

$\Longrightarrow$

| $a_1$ | $a_2$ | $a_3$ | $a_0$ |

**Fig. 2.** The **RotWord** operation on $\mathcal{AES}$

while the **xor** operation corresponds to the sum in $\mathbb{F}$. The $r^{\text{th}}$ round for the $\mathcal{AES}$ key generation scheme is the following one:

$$\mathcal{K}_A = \begin{cases} \mathbf{H}_{r0} = \varphi_A^r(\mathbf{H}_{r-1,3}) \\ \mathbf{H}_{r1} = \mathbf{H}_{r0} + \mathbf{H}_{r-1,1} \\ \mathbf{H}_{r2} = \mathbf{H}_{r1} + \mathbf{H}_{r-1,2} \\ \mathbf{H}_{r3} = \mathbf{H}_{r2} + \mathbf{H}_{r-1,3} \end{cases} \implies \mathbf{H}_r = (\mathbf{H}_{r0}, \mathbf{H}_{r1}, \mathbf{H}_{r2}, \mathbf{H}_{r3}) \qquad (4)$$

## 3   An Overview on the Big Encryption System ($\mathcal{BES}$)

Our starting point is the $\mathcal{BES}$ cipher, in which the $\mathcal{AES}$ is embedded by a "natural" mapping. The $\mathcal{BES}$ operations involve no computations in $GF(2)$, only in $\mathbb{F}$. This allows us to describe $\mathcal{AES}$ by means of polynomial equation systems. Solving the systems means to find the key or an alias of its, and therefore to break the code.

The state spaces of $\mathcal{AES}$ and $\mathcal{BES}$ are respectively $\mathbf{A} = \mathbb{F}^{16}$ and $\mathbf{B} = \mathbb{F}^{128}$. The basic tool for the embedding is the *conjugation* operation $\phi$, that considers for each value in $\mathbb{F}$ the vector of its successive square powers.

$$\mathbb{F} \ni a \longmapsto \phi(a) = \tilde{a} = (a^{2^0}, a^{2^1}, ..., a^{2^7}) \in \mathbb{F}^8 \qquad (5)$$
$$\mathbb{F}^n \ni \mathbf{a} \longmapsto \phi(\mathbf{a}) = \tilde{\mathbf{a}} = (\phi(a_0), ..., \phi(a_7)) \in \mathbb{F}^{8n} \qquad (6)$$

Thanks to the easy-to-verify properties (with $0^{-1} = 0$) :

$$\phi(\mathbf{a} + \mathbf{a}') = \phi(\mathbf{a}) + \phi(\mathbf{a}') \qquad \text{and} \qquad \phi(\mathbf{a}^{-1}) = \phi(\mathbf{a})^{-1}, \qquad (7)$$

we can put a one-to-one correspondence between $\mathcal{AES}$ and $\mathcal{BES}$ operations:

$$\mathbf{B_A} = \phi(\mathbf{A}) \subset \mathbf{B} \qquad (8)$$

as the subset of $\mathbf{B}$ corresponding to $\mathbf{A}$.

Let $\mathbf{p}, \mathbf{c} \in \mathbf{B}$ be the plaintext and ciphertext, respectively; $\mathbf{w}_i, \mathbf{x}_i \in \mathbf{B}$ ($0 \le i \le 9$) the mapped state vectors before and after the inversion phases that occur in the codifying process, and $\mathbf{h}_i \in \mathbf{B}$ the used keys.

All the phases of Rijndael algorithm may be translated in $\mathbf{B}$ using just linear algebra in $\mathbb{F}$, apart from inversion, which is simply done component-wise, as follows.

The matrix $L_A : \mathbb{F} \simeq GF(2)^8 \to GF(2)^8 \simeq \mathbb{F}$ for the affine transformation for one byte in the $\mathcal{S}$-box phase can be represented by the polynomial function $f : \mathbb{F} \to \mathbb{F}$:

$$f(a) = \sum_{k=0}^{7} \lambda_k a^{2^k} \tag{9}$$

$$
\begin{aligned}
\lambda_0 &= t^2 + 1 & \lambda_4 &= t^7 + t^6 + t^5 + t^4 + t^2 \\
\lambda_1 &= t^3 + 1 & \lambda_5 &= 1 \\
\lambda_2 &= t^7 + t^6 + t^5 + t^4 + t^3 + 1 & \lambda_6 &= t^7 + t^5 + t^4 + t^2 + 1 \\
\lambda_3 &= t^5 + t^2 + 1 & \lambda_7 &= t^7 + t^3 + t^2 + t + 1
\end{aligned} \tag{10}
$$

Working in **B**, $L_B(a) = \phi(L_A(a)) = (f(a)^{2^0}, \dots, f(a)^{2^7})$ and we must compute the successive squares of $f$: this is accomplished by finite induction, with basic step:

$$(f(a))^2 = \left( \sum_{k=0}^{7} \lambda_k a^{2^k} \right)^2 = \sum_{k=0}^{7} \lambda_k^2 a^{2^k \cdot 2} = \sum_{k=0}^{7} \lambda_k^2 a^{2^{k+1}} \tag{11}$$

Simply speaking, it is sufficient to iteratively square and circularly shift ($a^{2^8} = a = a^{2^0}$) the coefficients. The resulting matrix, which we still indicate with $L_B$, is

$$L_B = [l_{ij}]_{i,j=0,\dots 7} \qquad \text{with} \qquad l_{ij} = \lambda^{2^i}_{(8-i+j) \bmod 8} \tag{12}$$

The global transformation $\mathrm{Lin}_B : \mathbb{F}^{128} \to \mathbb{F}^{128}$ is the block diagonal matrix with 16 blocks equal to $L_B$.

The $\mathcal{AES}$ constant $c_A = 63 = t^6 + t^5 + t + 1 \in \mathbb{F}$ used in the $\mathcal{S}$-box maps into:

$$
\begin{aligned}
\phi(c_A) &= (63, \mathsf{C2}, 35, 66, \mathsf{D3}, \mathsf{2F}, 39, 36) \\
&= (t^6 + t^5 + t + 1, t^7 + t^6 + t, t^5 + t^4 + t^2 + 1, t^6 + t^5 + t^2 + t, \\
&\quad\ t^7 + t^6 + t^4 + t + 1, t^5 + t^3 + t^2 + t + 1, t^5 + t^4 + t^3 + 1, \\
&\quad\ t^5 + t^4 + t^2 + t)
\end{aligned} \tag{13}
$$

The corresponding $\mathcal{BES}$ vector $\mathbf{c}_B$ is simply obtained by the juxtaposition of 16 consecutive copies of $\phi(c)$, such that:

$$\mathbf{c}_B = \phi(\underbrace{c_A, \dots, c_A}_{16}) = (\underbrace{\phi(c_A), \dots, \phi(c_A)}_{16}) \qquad [\mathbf{c}_B]_i = [\phi(c_A)]_{i \bmod 8} \tag{14}$$

The $\mathcal{AES}$ **ShiftRows** transformation is given by the matrix $R_A : \mathbb{F}^{16} \to \mathbb{F}^{16}$:

$$
R_A = \left(\begin{array}{cccc|cccc|cccc|cccc}
1&0&0&0&0&0&0&0&0&0&0&0&0&0&0&0\\
0&0&0&0&0&1&0&0&0&0&0&0&0&0&0&0\\
0&0&0&0&0&0&0&0&0&0&1&0&0&0&0&0\\
0&0&0&0&0&0&0&0&0&0&0&0&0&0&0&1\\
0&0&0&0&1&0&0&0&0&0&0&0&0&0&0&0\\
0&0&0&0&0&0&0&0&0&1&0&0&0&0&0&0\\
0&0&0&0&0&0&0&0&0&0&0&0&0&0&1&0\\
0&0&0&1&0&0&0&0&0&0&0&0&0&0&0&0\\
0&0&0&0&0&0&0&1&0&0&0&0&0&0&0&0\\
0&0&0&0&0&0&0&0&0&0&0&0&0&1&0&0\\
0&0&1&0&0&0&0&0&0&0&0&0&0&0&0&0\\
0&0&0&0&0&0&0&1&0&0&0&0&0&0&0&0\\
0&0&0&0&0&0&0&0&0&0&0&0&1&0&0&0\\
0&1&0&0&0&0&0&0&0&0&0&0&0&0&0&0\\
0&0&0&0&0&0&1&0&0&0&0&0&0&0&0&0\\
0&0&0&0&0&0&0&0&0&0&0&1&0&0&0&0
\end{array}\right)
\tag{15}
$$

The corresponding $\mathcal{BES}$ matrix is obtained simply "expanding" each one in $R_A$ with an identity matrix of order 8, $I_8$, and each 0 with a zero $(8 \times 8)$ matrix. The result is $R_B : \mathbb{F}^{128} \to \mathbb{F}^{128}$.

The $\mathcal{AES}$ **MixColumns** may be represented using the $C_A : \mathbb{F}^4 \to \mathbb{F}^4$ matrix:

$$
C_A = \begin{pmatrix}
t & t+1 & 1 & 1 \\
1 & t & t+1 & 1 \\
1 & 1 & t & t+1 \\
t+1 & 1 & 1 & t
\end{pmatrix}
\tag{16}
$$

The $\mathcal{AES}$ transformation is given by the $\mathrm{Mix}_A : \mathbb{F}^{16} \to \mathbb{F}^{16}$ block diagonal matrix having as blocks four copies of $C_A$. In order to obtain the corresponding matrix we first need to compute the following matrices $C_B^{(k)}$, for $k = 0, \dots, 7$:

$$
C_B^{(k)} = \begin{pmatrix}
t^{2^k} & (t+1)^{2^k} & 1 & 1 \\
1 & t^{2^k} & (t+1)^{2^k} & 1 \\
1 & 1 & t^{2^k} & (t+1)^{2^k} \\
(t+1)^{2^k} & 1 & 1 & t^{2^k}
\end{pmatrix}
\tag{17}
$$

with:

$$
\begin{array}{lll}
t^{2^0} = t & t^{2^3} = t^4 + t^3 + t + 1 & t^{2^6} = t^6 + t^3 + t^2 + 1 \\
t^{2^1} = t^2 & t^{2^4} = t^6 + t^4 + t^3 + t^2 + t & t^{2^7} = t^7 + t^6 + t^5 + t^4 + t^3 + t \\
t^{2^2} = t^4 & t^{2^5} = t^7 + t^6 + t^5 + t^2 &
\end{array}
\tag{18}
$$

from which $(t+1)^{2^k} = t^{2^k} + 1$ may be very easily computed.

Using an appropriate basis (see below), the resulting matrix $M_B : \mathbb{F}^{128} \to \mathbb{F}^{128}$ may be written as a block diagonal one, having as blocks for all possible $k$

four consecutive copies of $C_B^{(k)}$. The change of basis is necessary because of the different positioning of value powers in $\phi$'s image with respect to what we need. Indeed, if $\mathbf{a} \in \mathbb{F}^{16}$, then:

$$\phi(\mathbf{a}) = (a_0, ..., a_0^{2^7}, a_1, ..., a_1^{2^7}, ..., a_{15}, ..., a_{15}^{2^7}) \qquad (19)$$

while in order to use the block diagonal representation, we would need the vector rearranged in this way:

$$\mathbf{a}' = (a_0, ..., a_{15}, a_0^2, ..., a_{15}^2, ..., a_0^{2^7}, ..., a_{15}^{2^7}) \qquad (20)$$

The corresponding permutation matrix $\mathrm{Perm}_B : \mathbb{F}^{128} \rightarrow \mathbb{F}^{128}$ permits to do this transformation. To write it down easily, suppose to divide it into $(16 \times 8)$ sub-matrices, called $P_{hk}$, $h = 0, \ldots, 7$, $k = 0, \ldots, 15$. Each sub-matrix element (with $i = 0, \ldots, 15$, $j = 0, \ldots, 7$) is thus defined:

$$[P_{hk}]_{ij} = \begin{cases} 1 \text{ if } i = k \text{ and } j = h \\ 0 \text{ else} \end{cases} \qquad (21)$$

Its inverse matrix $\mathrm{Perm}^{(-1)}$ is equally easy to describe: viewing it as composed of $(8 \times 16)$ sub-matrices $P_{hk}^{(-1)}$, with $h = 0, \ldots, 15$, $k = 0, \ldots, 7$, the generic element $[P_{hk}^{(-1)}]_{ij}$ (with $i = 0, \ldots, 7$, $j = 0, \ldots, 15$) is defined exactly as $[P_{hk}]_{ij}$ is. We therefore have $\mathrm{Mix}_B = \mathrm{Perm}_B^{-1} \cdot M_B \cdot \mathrm{Perm}_B$.

It is possible to avoid using $c_A$ slightly modifying the key generation scheme with respect to the original proposal. We show here how. If $\mathbf{b}, (\mathbf{h}_B)_i \in \mathbf{B}$ are the state and key vectors for the generic $i^{\mathrm{th}}$ round of $\mathcal{BES}$ associated to the corresponding one in $\mathcal{AES}$, we have:

$$\begin{aligned} \mathrm{Round}_B(\mathbf{b}, (\mathbf{h}_B)_i) &= \mathrm{Mix}_B(R_B(\mathrm{Lin}_B(\mathbf{b}^{-1}) + \mathbf{c}_B)) + (\mathbf{h}_B)_i \\ &= M_B \cdot (\mathbf{b}^{-1}) + (C_B(\mathbf{c}_B) + (\mathbf{h}_B)_i) \qquad (22) \\ &= M_B \cdot (\mathbf{b}^{-1}) + (\mathbf{k}_B)_i \end{aligned}$$

with:

$$M_B = \mathrm{Mix}_B \cdot R_B \cdot \mathrm{Lin}_B, \quad C_B = \mathrm{Mix}_B \cdot R_B, \quad (\mathbf{k}_B)_i = C_B(\mathbf{c}_B) + (\mathbf{h}_B)_i \quad (23)$$

The change on key generation scheme consists in adding a constant vector to each obtained round key, and this will be the form of the system we will work with.

### 3.1   $\mathcal{BES}$ Key Schedule Translation

- The $\mathcal{AES}$ **RotWord** operation is represented by a matrix $RW_A : \mathbb{F}^4 \longrightarrow \mathbb{F}^4$.

$$RW_A = \begin{pmatrix} 0 & 1 & 0 & 0 \\ 0 & 0 & 1 & 0 \\ 0 & 0 & 0 & 1 \\ 1 & 0 & 0 & 0 \end{pmatrix} \qquad (24)$$

The $\mathcal{BES}$ version $RW_B : \mathbb{F}^{32} \longrightarrow \mathbb{F}^{32}$ is obtained by replacing the ones with the identity matrix $I_8$, and the zeroes with the $(8 \times 8)$ zero matrix.

- The $\mathcal{S}$-box is applied only to a part of the whole vector, and therefore the corresponding matrix dimension changes. The resulting block diagonal matrix is $\mathrm{Lin}_B^k : \mathbb{F}^{32} \longrightarrow \mathbb{F}^{32}$, with four blocks equal to $L_B$.
- The constant $\mathbf{c}_B^k$ is obtained with just four copies of $\phi(c_A)$:

$$\mathbf{c}_B^k = \phi(c_A, c_A, c_A, c_A) = (\phi(c_A), \phi(c_A), \phi(c_A), \phi(c_A)), \quad [\mathbf{c}_B^k]_i = [\phi(c_A)]_{i \bmod 8} \tag{25}$$

- The constant vectors $\mathbf{Rcon}_i = (t^{i-1}, 0, 0, 0)$ are mapped into:

$$(\mathbf{Rcon}_B)_i = \phi(\mathbf{Rcon}_i) = (\phi(t^{r-1} =), \underbrace{0, \ldots, 0}_{24}) \tag{26}$$

In order to compute them, we need the $t^j$ normal forms with respect to $m(t)$.

To write in a very compact way the round key equations, we keep using the matrix notation, but in a *functional* sense. It is not possible to avoid here the use of constants.

If $\varphi_B^i : \mathbb{F}^{32} \to \mathbb{F}^{32}$ is the $\mathcal{BES}$ $i$th-round mapping function for a conjugated word $\mathbf{x}$:

$$\varphi_B^i(\mathbf{x}) = \mathrm{Lin}_B^k(RW_B(\mathbf{x}))^{-1} + \mathbf{c}_B^k + (\mathbf{Rcon}_B)_i \tag{27}$$

the generic $\mathcal{AES}$ and $\mathcal{BES}$ round matrices are $MK_A^i$ and $MK_B^i$:

$$MK_A^i = \begin{pmatrix} 0 & 0 & 0 & \varphi_A^i \\ 0 & I_4 & 0 & \varphi_A^i \\ 0 & I_4 & I_4 & \varphi_A^i \\ 0 & I_4 & I_4 & I_4 + \varphi_A^i \end{pmatrix} \text{ and } MK_B^i = \begin{pmatrix} 0 & 0 & 0 & \varphi_B^i \\ 0 & I_{32} & 0 & \varphi_B^i \\ 0 & I_{32} & I_{32} & \varphi_B^i \\ 0 & I_{32} & I_{32} & I_{32} + \varphi_B^i \end{pmatrix} \tag{28}$$

A key round is given by the computation of $\mathbf{h}_i = MK_B^i(\mathbf{h}_{r-1})$.

## 4    The Systems

We indicate how the processes for encryption and key generation can be represented by algebraic systems. We underline here once and for all that all the indicated variables satisfy the $\mathbb{F}$-belonging equation $y^{256} + y = 0$.

### 4.1    Encryption

Remembering that the last round differs slightly from the other ones, with $M_B^* = R_B \cdot \mathrm{Lin}_B$, the resulting system for codification is [8]:

$$\begin{cases} \mathbf{w}_0 = \mathbf{p} + \mathbf{k}_0 \\ \mathbf{x}_i = \mathbf{w}_i^{-1} & i = 0, \ldots, 9 \\ \mathbf{w}_i = M_B \mathbf{x}_{i-1} + \mathbf{k}_i & i = 1, \ldots, 9 \\ \mathbf{c} = M_B^* \mathbf{x}_9 + \mathbf{k}_{10} \end{cases} \tag{29}$$

Let here and in the rest of this paper the $(8j + m)^{\text{th}}$ component of all the vectors be indicated using the indexes expression $(j, m)$, with $j = 0, \ldots, 15$ and

$m = 0, \ldots, 7$. Under the hypothesis that no 0-inversion occurs (true for the 53% of encryptions and 85% of 128-bit keys), it is possible to expand the above system as follows, for all possible values of $j$ and $m$

$$
\begin{cases}
0 = w_{0,(j,m)} + p_{(j,m)} + k_{0,(j,m)} \\
0 = x_{i,(j,m)} w_{i,(j,m)} + 1 & i = 0, \ldots, 9 \\
0 = w_{i,(j,m)} + (M_B \mathbf{x}_{i-1})_{(j,m)} + k_{i,(j,m)} & i = 1, \ldots, 9 \\
0 = c_{(j,m)} + (M_B^* \mathbf{x}_9)_{(j,m)} + k_{10,(j,m)}
\end{cases}
\tag{30}
$$

Let $\alpha, \beta \in \mathbb{F}$ be the generic coefficients of $M_B$ and $M_B^*$, respectively. Now, adding the fact that the above equations should be valid for the $\mathbf{B_A}$ subset, that is, the state vectors must have the conjugation property, we finally have (with $m + 1$ considered mod 8):

$$
S = \begin{cases}
0 = w_{0,(j,m)} + p_{(j,m)} + k_{0,(j,m)} \\
0 = w_{i,(j,m)} + k_{i,(j,m)} + \displaystyle\sum_{(j',m')} \alpha_{(j,m),(j',m')} x_{i-1,(j',m')} & i = 1, \ldots, 9 \\
0 = c_{(j,m)} + k_{10,(j,m)} + \displaystyle\sum_{(j',m')} \beta_{(j,m),(j',m')} x_{9,(j',m')} \\
0 = x_{i,(j,m)} w_{i,(j,m)} + 1 & i = 0, \ldots, 9 \\
0 = x_{i,(j,m)}^2 + x_{i,(j,m+1)} & i = 0, \ldots, 9 \\
0 = w_{i,(j,m)}^2 + w_{i,(j,m+1)} & i = 0, \ldots, 9
\end{cases}
\tag{31}
$$

Let $S_\ell$, $\ell = 1, \ldots, 6$ indicate the equations in the $\ell^{\text{th}}$ line of the above system for all the possible values of $i$, $j$ and $m$, and $I_\ell$ the ideal they generate. As we see, the system is very sparse, with $S' = \{S_1, S_2, S_3\}$ linear, and the remaining equations in $S'' = \{S_4, S_5, S_6\}$ quadratic.

With $\mathbf{k} = \{k_i\}$, $\mathbf{w} = \{w_i\}$, $\mathbf{x} = \{x_i\}$, the numbers of the system are given in the following tables.

| Line | Number of equations | |
|------|---------------------|------|
| $S_1$ | $16 \cdot 8 =$ | 128 |
| $S_2$ | $9 \cdot 16 \cdot 8 =$ | 1152 |
| $S_3$ | $16 \cdot 8 =$ | 128 |
| $S_4$ | $10 \cdot 16 \cdot 8 =$ | 1280 |
| $S_5$ | $10 \cdot 16 \cdot 8 =$ | 1280 |
| $S_6$ | $10 \cdot 16 \cdot 8 =$ | 1280 |
| $S$ | Total $=$ | 5248 |

| Block | Number of variables | |
|-------|---------------------|------|
| $\mathbf{k}$ | $11 \cdot 16 \cdot 8 =$ | 1408 |
| $\mathbf{x}$ | $10 \cdot 16 \cdot 8 =$ | 1280 |
| $\mathbf{w}$ | $10 \cdot 16 \cdot 8 =$ | 1280 |
| | Total $=$ | 3968 |

## 4.2   Key Generation

It is possible to write down an analogous system for the key generation. It is more convenient to translate directly the $\mathcal{AES}$ procedure $\mathcal{K}_A$ into its $\mathcal{BES}$ counterpart $\mathcal{K}_B$, without explicitly expanding all the equations, as it is done in $MK_B^i$. The equations express all the $h_{i,(j,m)}$ variables in term of the $h_{0,(j,m)}$ ones. The index ranges for the equations are: $i = 1, \ldots, 10$, $\tilde{\jmath}, \tilde{\jmath}' = 0, \ldots, 3$ and $m, m' = 0, \ldots, 7$, while the $\text{Lin}_B^k$ matrix coefficients are indicated with $\gamma$.

$$\mathcal{K}_B = \begin{cases} \tilde{\mathbf{H}}_{i0} = \varphi_B^i(\tilde{\mathbf{H}}_{i-1,3}) \\ \tilde{\mathbf{H}}_{i1} = \tilde{\mathbf{H}}_{i0} + \tilde{\mathbf{H}}_{i-1,1} \\ \tilde{\mathbf{H}}_{i2} = \tilde{\mathbf{H}}_{i1} + \tilde{\mathbf{H}}_{i-1,2} \\ \tilde{\mathbf{H}}_{i3} = \tilde{\mathbf{H}}_{i2} + \tilde{\mathbf{H}}_{i-1,3} \end{cases}$$

$$= \begin{cases} z_{i,(\bar{\jmath},m)} = h_{i-1,(12+[(\bar{\jmath}+1) \bmod 4],m)}^{254} \\ h_{i,(\bar{\jmath},m)} = (\mathbf{c}_B^k + (\mathbf{Rcon}_B)_i)_{(\bar{\jmath},m)} + \sum_{(\bar{\jmath}',m')} \gamma_{(\bar{\jmath},m)(\bar{\jmath}',m')} z_{i,(\bar{\jmath}',m')} \\ h_{i,(4s+\bar{\jmath},m)} = h_{i,(4(s-1)+\bar{\jmath},m)} + h_{i-1,(4s+\bar{\jmath},m)} \qquad s = 1,2,3 \end{cases}$$

Let $\mathbf{cR}_i = \mathbf{c}_B^k + (\mathbf{Rcon}_B)_i$ be the constant vector occurring in each round, and its elements be indicated with $\delta_i$. Remembering the third equivalence of (23), with $t = 0, \ldots, 15$ and the conjugation property, we can obtain the keys with:

$$K = \begin{cases} 0 = z_{i,(\bar{\jmath},m)} + h_{i-1,(12+[(\bar{\jmath}+1) \bmod 4],m)}^{254} \\ 0 = h_{i,(\bar{\jmath},m)} + \delta_{i,(\bar{\jmath},m)} + \sum_{(\bar{\jmath}',m')} \gamma_{(\bar{\jmath},m)(\bar{\jmath}',m')} z_{i,(\bar{\jmath}',m')} \\ 0 = h_{i,(4s+\bar{\jmath},m)} + h_{i,(4(s-1)+\bar{\jmath},m)} + h_{i-1,(4s+\bar{\jmath},m)} \qquad s = 1,2,3 \\ \\ 0 = k_{i,(t,m)} + (C_B(\mathbf{c_B}))_{(t,m)} + h_{i,(t,m)} \\ 0 = z_{i,(\bar{\jmath},m)}^2 + z_{i,(\bar{\jmath},m+1)} \\ 0 = h_{i,(\bar{\jmath},m)}^2 + h_{i,(\bar{\jmath},m+1)} \end{cases}$$

$$(32)$$

## 5   System Solving

Actually, we are interested to recover the key out of the systems $S$ and $K$, that is the original key $\mathbf{h} = \phi^{-1}(\mathbf{k}^\star) = \{h_0, \ldots, h_{15}\}$, where $\mathbf{k}^\star = \{k_{0,(0,m)}, \ldots, k_{0,(15,m)}\}$. This is the task of this section.

In order to obtain equations relating $\mathbf{h}$ ($\mathbf{k}$) components we have to eliminate all other variables. We will do this by:

- modifying the way the systems are written,
- doing some variable substitutions "by hand" (see below), and finally
- perform Gröbner bases computations (more complicated substitutions, expansions and simplifications) in order to obtain the final system.

First of all, we write the systems in a more appropriate way for our purposes. Observe that, for each variable $v \in \mathbf{k}, \mathbf{w}, \mathbf{z}, \mathbf{h}$, the conjugation property of $\mathcal{BES}$ vectors may be synthesized by the obvious following relations:

$$v_{i,(j,m)} = v_{i,(j,0)}^{2^m} \qquad m = 0, \ldots, 7 \qquad (33)$$

### 5.1   Encryption

We rewrite $S$ as follows: first of all, we remove the imposed restriction about inversion, and we substitute $S_4$ with an equation expressing the true definition of the general inverse of an element of $\mathbb{F}$. Then we use (33), in order to remove all the variables with index $m > 0$, and obtain:

$$S^* = \begin{cases} 0 = w_{0,(j,0)}^{2^m} + p_{(j,0)}^{2^m} + k_{0,(j,0)}^{2^m} \\[2mm] 0 = w_{i,(j,0)}^{2^m} + k_{i,(j,0)}^{2^m} + \displaystyle\sum_{(j',m')} \alpha_{(j,m),(j',m')} x_{i-1,(j',0)}^{2^{m'}} \quad i = 1,\ldots,9 \\[2mm] 0 = c_{(j,0)}^{2^m} + k_{10,(j,0)}^{2^m} + \displaystyle\sum_{(j',m')} \beta_{(j,m),(j',m')} x_{9,(j',0)}^{2^{m'}} \\[2mm] 0 = x_{i,(j,0)} + w_{i,(j,0)}^{254} \quad\quad\quad\quad i = 0,\ldots,9 \end{cases} \quad (34)$$

We use the last equation to remove all the $x_{i,(j,0)}$, and, being each line a set of successive square powers, we keep only the ones with $m = 0$, obtaining:

$$S^* = \begin{cases} 0 = w_{0,(j,0)} + p_{(j,0)} + k_{0,(j,0)} \\[2mm] 0 = w_{i,(j,0)} + k_{i,(j,0)} + \displaystyle\sum_{(j',m')} \alpha_{(j,0),(j',m')} w_{i-1,(j',0)}^{254\cdot 2^{m'}} \quad i = 1,\ldots,9 \\[2mm] 0 = c_{(j,0)} + k_{10,(j,0)} + \displaystyle\sum_{(j',m')} \beta_{(j,0),(j',m')} w_{9,(j',0)}^{254\cdot 2^{m'}} \end{cases} \quad (35)$$

Because of the block structure of matrices $\text{Lin}_B$ and $R_B$, the $\beta$ coefficients do not depend on $j$ and $j'$, and the values are simply the coefficients of $f$. To further simplify the notations, hereafter we take, $\mod 255$:

$$\omega = (\omega_0,\ldots,\omega_7) = (254\cdot 2^0,\ldots,254\cdot 2^7) = (254,253,251,247,239,223,191,127),$$
$$\omega' = (\omega'_0,\ldots,\omega'_7) = (\omega_0 - 127,\ldots,\omega_7 - 127) = (127,126,124,120,112,96,64,0) .$$

be two auxiliary vectors. We can now avoid writing $m$ index, and finally write:

$$S^* = \begin{cases} 0 = w_{0,j} + p_j + k_{0,j} \\[2mm] 0 = w_{i,j} + k_{i,j} + \displaystyle\sum_{(j',m')} \alpha_{(j,0),(j',m')} w_{i-1,j'}^{\omega_{m'}} \quad i = 1,\ldots,9 \\[2mm] 0 = k_{10,j} + c_j + \displaystyle\sum_{m'} \lambda_{m'} w_{9,j'}^{\omega_{m'}} \end{cases} \quad (36)$$

The system so obtained has $16 + 9\cdot 16 + 16 = 176$ equations in $11\cdot 16 + 10\cdot 16 = 336$ variables. Obviously, it expresses nothing but a series of successive substitutions, down to the last equation. If we consider the block lex order for which, independently from the lex order inside each block,

$$\mathbf{k_{10}} > \mathbf{w_9} > \mathbf{k_9} > \cdots > \mathbf{w_0} > \mathbf{k_0} \quad (37)$$

we have a (not reduced) Gröbner basis [1], and the above substitutions may be considered as the complete reduction computation. The resulting set of the last 16 equations, where all the $\mathbf{w}$ variables disappeared, is what we were looking for. Indicating with $q_j^S$ the polynomials resulting from the substitutions, we have:

$$k_{10,j} + c_j + q_j^S(\mathbf{k_0},\ldots,\mathbf{k_9},p) = 0 \quad\quad j = 0,\ldots,15 \quad (38)$$

## 5.2    Key Generation

We now try to get more informations analyzing $K$. We may:

- substitute **z** variables in the second line equations, to get rid of them.
- use the conjugation property, expressing everything in terms of variables with $m = 0$.
- note that the $C_B(\mathbf{c_B})$ constant vector has $c_A = t^6 + t^5 + t + 1$ in each of its $(j, 0)$ positions, and opportune powers in the other ones. In other words, the equations on the fourth line of $K$, $K_4$, may be reduced (the other ones are powers of it) to:

$$h_{i,(j,0)} + k_{i,(j,0)} + c_A = 0 \tag{39}$$

- for the above considerations, express everything directly in term of **k** variables.
- observe that $\mathrm{Lin}_B^k$ is a block diagonal matrix, and therefore just $\tilde{j}' = \tilde{j}$ is "active" for each single equation, and what remains is nothing more than the set of coefficients of the $f$ polynomial.

After the elaboration and always keeping present that we work in $\mathbb{F}$, the modified key generation scheme translates into the following system (where $i = 1, \ldots, 10$; $s, \tilde{j} = 0, \ldots, 3$ and in the las= t version we omit $m$). For brevity, we define the function:

$$\mathrm{in} : \mathbb{N} \ni n \to \mathrm{in}(n) = 12 + [(n+1) \mod 4] \in \mathbb{N} \tag{40}$$

$$K^\star = \begin{cases} 0 = h_{i,(\tilde{j},0)} + \delta_{i,(\tilde{j},0)} + \displaystyle\sum_{(\tilde{j}',m')} \gamma_{(\tilde{j},0)(\tilde{j}',m')} h_{i-1,(\mathrm{in}(\tilde{j}'),0)}^{254\cdot 2^{m'}} \\ 0 = h_{i,(4s+\tilde{j},0)} + h_{i,(4(s-1)+\tilde{j},0)} + h_{i-1,(4s+\tilde{j},0)} \\ 0 = h_{i,(\tilde{j},0)} + (k_{i,(\tilde{j},0)} + c_A) \end{cases}$$

$$= \begin{cases} 0 = (k_{i,(\tilde{j},0)} + c_A) + \delta_{i,(\tilde{j},0)} + \displaystyle\sum_{m'} \gamma_{(\tilde{j},0)(\tilde{j},m')} \left(k_{i-1,(\mathrm{in}(\tilde{j}),0)} + c_A\right)^{\omega_{m'}} \\ 0 = \left(k_{i,(4s+\tilde{j},0)} + c_A\right) + \left(k_{i,(4(s-1)+\tilde{j},0)} + c_A\right) + \left(k_{i-1,(4s+\tilde{j},0)} + c_A\right) \end{cases}$$

$$= \begin{cases} 0 = k_{i,\tilde{j}} + \left(c_A + \delta_{i,(\tilde{j},0)}\right) \\ \quad + \left(k_{i-1,\mathrm{in}(\tilde{j})} + c_A\right)^{127} \cdot \left(\displaystyle\sum_{m'} \lambda_{m'}(k_{i-1,\mathrm{in}(\tilde{j})} + c_A)^{\omega'_{m'}}\right) \\ 0 = k_{i,4s+\tilde{j}} + k_{i,4(s-1)+\tilde{j}} + k_{i-1,4s+\tilde{j}} + c_A \end{cases}$$

Now only **k** variables remain, 160 equations in 176 variables, and by successive substitutions we can express all the ones with $i > 0$ as polynomials in the "parameters" $\mathbf{k_0}$. The equations are a Gröbner basis for several suitable lex orderings. Its complete reduction may be computed considering, for example, the one such that

$$k_{10,15} > \cdots > k_{10,0} > \cdots > k_{0,15} > \cdots > k_{0,0} \tag{41}$$

Now we can do backward substitutions, à la Gauss. Note that it is possible to work with $h$ variables in order to have the equations following the original $\mathcal{AES}$

definition, and use (39) only at the end, in order to obtain the modified key generation scheme. In any case, the result is:

$$k_{i,j} = q_{i,j}^K(\mathbf{k_0}) \qquad i = 1, \ldots, 10 \quad , \quad j = 0, \ldots, 15 \qquad (42)$$

The final step consists in putting together the results obtained in the former sections. Now we have two main possibilities, depending on the number of $(p, c)$ pairs we may use.

**A single $(p, c)$ pair.** We have to eliminate all the generated intermediate keys, putting together the systems $S^\star$ and $K^\star$, with the refinement of (37) suggested by (41). In this way we obtain the entire substitution process once and for all. It is summarized by the insertion of (42) into (38):

$$\mathcal{C}_1 = \{ q_{10,j}^K(\mathbf{k_0}) + c_j + q_j^S\left(\mathbf{k_0}, q_{1,j}^K(\mathbf{k_0}), \ldots, q_{9,j}^K(\mathbf{k_0}), p\right) = 0 \quad | \quad j = 0, \ldots, 15 \}$$
$$(43)$$

which is a system of 16 equations in 16 variables, whose roots give the desired keys.

**Eleven or more $(p, c)$ pairs.** We may simply use a copy of (38) for each $(p, c)$ pair, in order to obtain a system in 176 variables with 176 or more equations, whose roots give *all* the desired keys, too.

$$\mathcal{C}_2 = \{ k_{10,j} + c_j^{(n)} + q_j^S\left(\mathbf{k_0}, \ldots, \mathbf{k_9}, p^{(n)}\right) = 0 \mid n = 1, \ldots, d \,, \, j = 0, \ldots, 15 \}$$
$$(44)$$

These systems are extremely dense, and a very big computation power is required to solve them. Obviously, making use of more $(p, c)$ pairs, we may render the system to be solved overdetermined. We may use these tools jointly, and so on.

## 6    Conclusions

We presented some algebraic aspects of representing $\mathcal{AES}$ as a system of polynomial equations following the $\mathcal{BES}$ approach. By means of successive substitutions, we were able to eliminate all the intermediate variables, and obtain two systems $S^\star$ and $K^\star$ whose solution corresponds exactly to code breaking. Actually, they are rather complicated. Solving them is not trivial at all.

$K^\star$ and most of $S^\star$ are invariant for all choices of keys. When extended, their joint size is of about 500 Kb. Each of them is a (not reduced) Gröbner basis for several lex orderings, their union is not. Probably there exists some ordering for which the calculus of a Gröbner basis is easier. If we ever can obtain this with reasonable computational resources, then $\mathcal{AES}$ can be declared broken.

Succeeding to calculate the Hilbert series of $K^\star \cup S^\star$, we should easily obtain the number $n_s$ of its solutions. We suspect that $n_s$ is invariant for all key and $(p, c)$ choices. Furthermore, we expect that $n_s$ expresses the redundancy of the keyspace of $\mathcal{AES}$. That is, it tells us how many key choices will set up the same bijection between the cleartext space and ciphertext space. The number of such bijections is expected to be:

$$\frac{\#(\mathcal{AES} \text{ Keyspace})}{n_s} \tag{45}$$

Probably a reasonably simple canonical representation of such bijections can be found. In this case, if $n_s$ is big enough, probably the right (unique up to the isomorphism) key can be found using exhaustive search.

# References

1. D. A. Cox, J. Little, D. O'Shea. Ideals, Varieties, and Algorithms, An Introduction to Computational Algebraic Geometry and Commutative Algebra. Springer-Verlag, New York, 1992.
2. N. Courtois, J. Pieprzyk. Cryptanalysis of block ciphers with overdefined systems of equations. IACR eprint server www.iacr.org, April 2002.
3. J. Daemen, V. Rijmen. AES proposal: Rijndael (Version 2). NIST AES website: csrc.nist.gov/encryption/aes, 1999.
4. J. Daemen, V. Rijmen. The design of Rijndael: AES - The Advanced Encryption Standard. Springer-Verlag, 2002.
5. National Institute of Standards and Technology. Advanced Encryption Standard. FIPS 197. 26 November 2001.
6. N. Ferguson, R. Schroeppel, D. Whiting. A simple algebraic representation of Rijndael. In *Selected Areas in Cryptography*, Proc. SAC 2001, Lecture Notes in Computer Science 2259, pp. 103-111, Springer Verlag, 2001.
7. Grayson, Daniel R. and Stillman, Michael E. Macaulay 2, a software system for research in algebraic geome= try. Available at http://www.math.uiuc.edu/Macaulay2/.
8. G.-M. Greuel, G. Pfister, H. Schönemann. SINGULAR 2-0-3. A Computer Algebra System for Polynomial Computations. Center for Computer Algebra, University of Kaiser slautern, 2003. www.singular.uni-kl.de.
9. S. Murphy M. J.B. Robshaw. *Essential Algebraic Structure within the AES*. M. Yung (ed.): CRYPTO 2002, LNCS 2242, pp. 1-16, Springer-Verlag 2002.
10. E. Oswald, J. Daemen, and V. Rijmen. *The State of the Art of Rijndael's Security*. Technical report. Available at www.a-sit.at/technologieb/evaluation/aes_report_e.pdf.
11. D. R. Stinson. CRYPTOGRAPHY, *Theory and Practice*. Chapman & Hall/CRC, 2002. Second edition.

# Efficient AES Implementations on ASICs and FPGAs

Norbert Pramstaller, Stefan Mangard, Sandra Dominikus,
and Johannes Wolkerstorfer

Institute for Applied Information Processing and Communications (IAIK),
Graz University of Technology, Inffeldgasse 16a, A-8010 Graz, Austria
{firstname.surname}@iaik.at

**Abstract.** In this article, we present two AES hardware architectures: one for ASICs and one for FPGAs. Both architectures utilize the similarities of encryption and decryption to provide a high throughput using only a relatively small area. The presented architectures can be used in a wide range of applications. The architecture for ASIC implementations is suited for full-custom as well as for semi-custom design flows. The architecture for the FPGA implementation does not require on-chip block RAMs and can therefore even be used for low-cost FPGAs.

**Keywords:** Advanced Encryption Standard (AES), FPGA, ASIC.

## 1 Introduction

The symmetric block cipher Rijndael [4] has been standardized by NIST[1] as Advanced Encryption Standard (AES) [10] in November 2001. Today, AES is the most widely used symmetric block cipher and it is implemented in many different devices to secure wired as well as wireless connection links.

The requirements for an implementation of AES strongly depend on the application running on the device and of course also on the type of the device. In many scenarios, it is sufficient to implement AES in software. However, there are also many very relevant scenarios, where the requirements concerning the implementation of AES can only be met by dedicated hardware implementations.

Encryption engines for high speed communication links, for example, often need to be implemented in hardware due to the high throughput requirements. In applications that need to be resistant against side-channel attacks, AES is also often implemented in special hardware modules [11]. The reason for this is simply that it is easier to secure a small AES module against side-channel attacks rather than to secure an entire processor. In devices where the power consumption is critical, hardware implementations are also the preferred choice, because they consume considerably less power than software implementations.

---

[1] National Institute of Standards and Technology.

H. Dobbertin, V. Rijmen, A. Sowa (Eds.): AES 2004, LNCS 3373, pp. 98–112, 2005.
© Springer-Verlag Berlin Heidelberg 2005

In this article, we present efficient hardware implementations for ASICs and FPGAs that can be used for a wide range of applications. Both architectures use similarities of encryption and decryption to provide a high level of performance using only a relatively small area. While most publications on implementations of AES only provide performance and area figures without interfaces and registers for CBC mode, the architectures presented in this article are complete. Both architectures include an AMBA APB [1] interface and can perform encryptions and decryptions in CBC mode.

Our architecture for ASICs is presented in Section 2 and the one for FPGAs is discussed in Section 3. Conclusions about both architectures can be found in Section 4.

## 2   ASIC Implementation of AES

During the last years, several proposals [8, 12, 13, 16, 19] on how to implement AES on an ASIC have been published. Most of these publications focus mainly on the throughput of the implementation. In this article however, we describe an architecture that in addition is highly regular and scalable. Therefore, it can be used for a wide range of applications.

The architecture discussed in this article has balanced combinational paths in order to fully utilize every clock cycle. The fact that the combinational paths are short compared to other published AES architectures, makes the presented architecture a favorable choice for low-power applications. This is due to the fact that glitches, which occur more frequently in long combinational paths than in short ones, cause a significant power consumption.

The regularity of the presented architecture helps to keep the size of the AES architecture small during place-and-route of a semi-custom design flow and facilitates the creation of full-custom designs. Full-custom approaches are interesting for smart card implementations that are required to provide protection against power analysis attacks [7]. In a full-custom approach, the designer can well balance the capacitive loads of output nodes, as it is for example desired for logic styles like the one described in [14, 15].

The performance of our AES architecture can gradually be increased at the cost of an increased chip size. The overall structure of this architecture, which is capable of performing AES encryptions and decryptions, is shown in Fig. 1. The AES hardware module consists of the following four components:

– **The Interface:** The AMBA APB interface handles all communication of the AES module with its environment.
– **The Data Unit:** The data unit is the main module of the architecture. It can perform any kind of AES encryption or decryption round using the round key that is assigned to its key input. Although the number of rounds is different for the three standardized key sizes, the types of rounds which need to be executed are the same for all key sizes. Consequently, the data unit is independent of the key size.

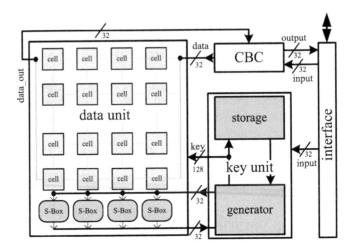

**Fig. 1.** Overall structure of the AES module

- **The Key Unit:** The key unit serves two main purposes: the storage of cipher keys and the calculation of the round keys. To save die size, the S-Boxes of the data unit are reused to perform the key expansion. In the presented architecture, this reuse is possible for any key size without loss of performance.
- **The CBC Unit:** The CBC unit of the AES module implements the CBC mode without any negative influence on the overall performance of the AES module.

In the presented architecture, a 128-bit block of data is encrypted as follows. First, a cipher key needs to be loaded via the AMBA APB interface into the key unit. Once a key is loaded, it can be used for an arbitrary number of encryptions and decryptions. After the loading of the cipher key, the first 128-bit block of data is transferred via the interface and the CBC unit into the data unit. The data unit then iteratively performs the number of AES rounds that are required for the used key size.

In each round, the key unit provides the corresponding round key to the data unit. To calculate these round keys, the key unit uses the S-Boxes of the data unit during a clock cycle in which they are not required by the data unit. After the calculation of the AES rounds, the encryption result is passed in 32-bit words to the interface via the CBC unit. Decryptions are computed in a very similar way. In this case, the data unit performs the inverse AES transformations in reversed order and also the key unit provides the round keys in reversed order.

The remainder of this section presents the details of the data and the key unit.

## 2.1   The Data Unit

The data unit is the biggest and the most important component of the AES architecture. It stores the current 128-bit data block of an encryption or de-

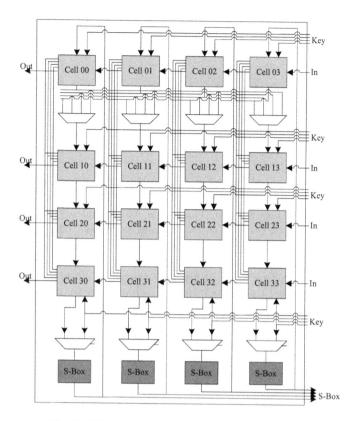

**Fig. 2.** The architecture of the standard data unit

cryption (referred to as the "State") and is capable of performing any number and type of en-/decryption rounds on this State. Consequently, all four AES transformations (SubBytes, ShiftRows, MixColumns and AddRoundKey) and the corresponding inverse transformations are implemented within the data unit. For the AddRoundKey transformation, a corresponding round key needs to be provided by the key unit.

Figure 2 shows the standard version of the data unit. It consists of sixteen so-called data cells and four S-Boxes. An S-Box of the architecture is a circuit capable of performing the SubBytes and the inverse SubBytes transformation for an 8-bit input. The data cells store eight bits per cell and perform all other AES transformations and the corresponding inverses, when connected appropriately. In full-custom designs, inputs and outputs of the data cells can be defined such that connection by abutment is possible when they are placed next to another.

Another distinguishing feature of the presented architecture is the fact that the combinational paths are relatively short and, more important, very balanced. The commonly used approach to implement AES in hardware is to store the 128-bit State in a register and to perform the AES transformations (except for the ShiftRows transformation) column by column. Therefore, to perform a normal

AES encryption round, first the ShiftRows transformation is done in one clock cycle. Then, the remaining transformations of an AES round are applied column by column, whereby all transformations for one column are usually done within one clock cycle.

The problem of this approach is that the combinational path to perform a SubBytes, a MixColumn and an AddRoundKey transformation in one clock cycle is very long. Additionally, the implementation of the ShiftRows transformation causes a significant wiring overhead. The data unit, presented in this section, solves both problems. It performs AES encryptions and decryptions in the following way:

To load a data block, the input data is shifted column by column from the right side (see Fig. 2) into the data cells. The inputs labelled "In" are connected via the CBC unit to the interface. The initial AddRoundKey transformation is done in the fourth clock cycle at the same time as the last column is loaded.

To compute a normal AES round, the registers are rotated vertically to perform the (Inv)SubBytes and the (Inv)ShiftRows transformation row by row. In our design, we use an S-Box with one pipeline stage. Therefore, the (Inv)SubBytes and the (Inv)ShiftRows transformations can be applied to all sixteen bytes of the State in five clock cycles.

In the sixth clock cycle of a normal AES round, the (Inv)MixColumns and the AddRoundKey transformations are performed by all data cells in parallel. Since the S-Boxes are not used by the data unit during the sixth clock cycle, they can be utilized by the key unit to perform the key expansion for the next round key.

This way, the required number of encryption or decryption rounds can be executed by the data unit and the key unit until the 128-bit result is finally stored in the registers of the data unit. This result is then shifted column by column to the left (to the interface of the AES module). At the same time, a new input State can be loaded.

Using the standard data unit, the minimal number of clock cycles that are required to perform an AES-128 encryption or decryption is 65. Four clock cycles are required for the I/O of the data unit, 54 clock cycles are required to perform the nine normal AES rounds and seven are required for the final round.

The following two subsections present the architecture of the S-Boxes and the data cells.

**S-Boxes.** In hardware implementations, the SubBytes transformation and its inverse are the most expensive AES transformations. For the presented AES module, a pipelined (one stage) implementation of the S-Box as described in [18] is used. The main idea of this implementation is to build an efficient combinational circuit for the S-Box, which is based on the fact that $GF(2^8)$ can be seen as quadratic extension of the field $GF(2^4)$. A pipelined version of the S-Box is used to accomplish that the combinational paths in the architecture are balanced (*i.e.* the paths of the S-Boxes and those of a MixColumns-and-AddRoundKey step are roughly the same).

**Data Cells.** The design of the data cells is crucial for the overall architecture of the data unit. The data cells serve as storage elements of the AES State and perform the (Inv)MixColumns and the AddRoundKey transformation. Besides some input selection circuit, each data cell consists of eight flip-flops, a multiplier [17] for the MixColums transformation (which can also be omitted in order to scale the design) and XOR gates for the AddRoundKey transformation.

## 2.2   The Key Unit

The key unit is used to store keys and to calculate the key expansion function. Due to the fact that the AES is standardized for 128, 192 and 256-bit keys, the interface between the key unit and the data unit is designed such that the key expansion for several different key sizes can be implemented on the same chip.

The key unit used in our design stores the key loaded via the interface and is capable of calculating all round keys for encryption and decryption iteratively. Since the data unit does not perform any S-Box lookups while the MixColumns and AddRoundKey transformations are executed, the S-Boxes of the data unit are reused by the key unit during this clock cycle. Details about the key unit can be found in [8].

## 2.3   Performance of the ASIC AES Implementation

As mentioned before, the presented AES module is built up very regular and is highly scalable. Three different scaled versions of the module have been implemented and tested. They are named "standard version", which was described in the previous sections, "minimum version", which is the smallest, but slowest one, and the "high-performance version", which is the fastest, but most area intensive version.

As shown in Fig. 2, the data unit of the standard version consists of 16 data cells (including 16 MixColumns multipliers) and four S-Boxes. The en-/decryption of an 128-bit data block requires 65 clock cycles. By using 16 S-Boxes instead of four, the performance can be increased. The S-Box lookup can be done for all 128 bits of the State in parallel. With this configuration (high-performance version), the AES module requires only 35 clock cycles to en-/decrypt a 128-bit data block. In the minimum version of the AES module, only four S-Boxes and four MixColumns multipliers are used. Only the four "leftmost" data cells contain MixColumns multipliers. In the other data cells the multipliers are omitted. Here, 92 clock cycles are needed to process a 128-bit data block.

In this section, we give a comparison of the three different scaled versions of the AES module in terms of performance and area. Additionally, a comparison to related work is given.

## 2.4   Performance of the Presented ASIC Design

The three versions of AES-128 have been implemented in VHDL and have been synthesized for a 0.6 $\mu m$ CMOS process. Table 1 shows the complexity in gate

**Table 1.** Complexity of AES components in GE

| Component | Complexity [GE] |
|---|---|
| S-Box | 392 |
| Multiplier | 212 |
| Data cell (without Multiplier) | 87 |
| Key generator | 1,633 |
| Key store | 691 |
| AMBA Bus Interface | 267 |
| CBC Register | 1,599 |

**Table 2.** Complexity of the AES-128 modules

| Component | # | Minimum | # | Standard | # | High Perf. |
|---|---|---|---|---|---|---|
| S-Boxes | 4 | 1,568 | 4 | 1,568 | 16 | 6,272 |
| Multipliers | 4 | 848 | 16 | 3,392 | 16 | 3,392 |
| D. cells without mult. | 16 | 1,392 | 16 | 1,392 | 16 | 1,392 |
| Multiplexors | 96 | 224 | 192 | 384 | 224 | 374 |
| **Data unit** | | **4,032** | | **6,736** | | **11,430** |
| Key generator | 1 | 1,633 | 1 | 1,633 | 1 | 1,633 |
| Key store | 1 | 691 | 1 | 691 | 1 | 691 |
| **Key unit** | | **2,324** | | **2,324** | | **2,324** |
| AMBA + CBC | 1 | 1,866 | 1 | 1,866 | 1 | 1,866 |
| Control logic | | 319 | | 279 | | 230 |
| **Additional** | | **2,185** | | **2,145** | | **2,096** |
| **Total** | | **8,541** | | **11,205** | | **15,850** |

equivalents (GE) of each component used for the AES module. In Table 2, the complexity of the three different modules is calculated by adding the size of the components used for the different versions. In the first column for each version the used number of components is given. The standard version of the module has a complexity of 11,205 GE, whereas the minimum version needs 8,541 GE. The high performance version requires 15,850 GE.

The high-performance module essentially consists of 12 S-Boxes more than the standard module. This causes an increase of the complexity by 41%. On the other hand, the minimum version consists of 12 MixColumns multipliers less than the standard version and is therefore 24% smaller. The critical path of all three designs is more or less the same and is determined by the delay of one pipeline stage of the S-Box—the maximum frequency for the complete AES-128 modules (including AMBA interface, CBC, and control logic) on the 0.6 $\mu m$ technology is about 50 MHz. In Table 3, a summary of the performance is shown.

The standard version needs 65, the high-performance version 35, and the minimum version 92 clock cycles to perform an AES-128 encryption or decryption. In the high-performance version, the improvement of the throughput by 87% is paid by an increase of the complexity by 41%. In the minimum version, the reduced complexity (-24%) accounts for a 29% loss in throughput.

**Table 3.** Summary of the performance of the different AES-128 modules

| Version | Clock Cycles | Throughput@50 MHz [$Mbps$] | Area [GE] |
|---|---|---|---|
| Minimum | 92 | 70 | 8,541 |
| Standard | 65 | 98 | 11,205 |
| High perf. | 35 | 183 | 15,850 |

## 2.5 Related Work

This subsection compares the presented architecture with the one proposed in [13]. The design of Satoh et al. consists of 5,400 GE and was implemented on a 0.11 $\mu m$ technology. The design consists of a core data path and a key generator. It does not include mechanisms for I/O, CBC registers or a key store. Its maximum clock frequency is about 130 MHz. The design requires 54 cycles to perform an encryption, which leads to a theoretical throughput (for the four-S-Box version) of 311 Mbps.

For a comparison of complexity, the gate count for our design has to be reduced by the additional components our design offers (key store, CBC registers, AMBA interface)—this leads to a gate count of about 8,600 GE for the standard AES-128 module. This comparison is still not completely fair, since different technologies are used for synthesis. The 0.11 $\mu m$ technology used in [13] allows smaller structures and offers different leaf cells. It seems to be more extensive than our technology, because similar components of the designs are claimed to be smaller in the architecture in [13]. For example, the S-Box proposed by Satoh et al. was reconstructed and synthesized with our technology. The result was a 15% bigger S-Box than used in our design, whereas the results with the 0.11 $\mu m$ technology in [13] show an S-Box that needs 25% less GE than our S-Box design in the 0.6 $\mu m$ technology.

The big difference in the used technology also does not allow a reasonable comparison of the maximum frequencies or the throughput. When comparing the two proposed S-Boxes synthesized in our 0.6 $\mu m$ technology in terms of delay, our proposed S-Box is about 30% faster. An attempt to compare the throughput of both designs starts with comparing the critical paths. In [13] the critical path is very long: The SubBytes, the MixColumns and the AddRoundKey transformation are done for one column within one clock cycle. Additionally, in the same clock cycle the data passes the so-called selector function, which seems to be another major cause of delay.

In our presented architecture, the critical path consists only of one pipeline stage of an S-Box. This is approximately a third of the critical path of the architecture presented in [13]. Using the same technology for synthesis of the compared designs, we expect the maximum frequency of our module to be at least three times higher than the maximum frequency stated in [13]. This leads to a better overall performance.

## 3    FPGA Implementation of AES

Reconfigurable devices like Field Programmable Logic Arrays (FPGA) gain more and more importance in hardware, software and hardware/software co-designs. In the beginning often seen only as devices for rapid prototyping, FPGAs are increasingly used for final applications. One of the mostly used arguments for using FPGAs rather than ASICs, is the reduced time-to-market and the cost advantages of standard devices.

Due to the importance of reconfigurable devices, numerous FPGA AES implementations have been published within the last years. These implementations mainly focus on high throughput rates [3, 9], and use techniques like loop unrolling and pipelining. They are able to report throughput rates up to 12,160 Mbps [3]. Applying such techniques leads to AES hardware implementations that require a huge amount of FPGA resources that are only available for expensive devices and cannot be implemented in low-end FPGAs.

In this section we present a new universal architecture that is supported by several FPGA product families and can also be implemented using inexpensive low-end FPGAs. It is the first known AES FPGA implementation that does not require on-chip block RAMs. Furthermore, it implements the complete AES encryption standard and features the Cipher Block Chaining (CBC) mode.

### 3.1    Related Work

Gaj et al. [3] published the fastest known FPGA implementation. For encryption and decryption with 128-bit keys, a throughput of 12,160 Mbps on a Xilinx Virtex XCV1000BG560-6 device is reported. McLoone et al. [9] achieve a throughput of 6,956 Mbps for 128-bit keys only. They also presented encryption engines for 192-bit or 256-bit keys with accordingly lower throughput. Their combined encryption and decryption implementation can handle 128-bit keys and achieves a throughput of 3,239 Mbps on a Xilinx Virtex-E XCV3200E-8CG1156 device. The third implementation published by Dandalis et al. [5] also provides encryption and decryption for 128-bit keys. They achieve a throughput of 353 Mbps on a Xilinx Virtex XCV1000BG560-6 device. Fischer et al. [6] published a non-pipelined design supporting encryption and decryption for 128-bit keys. They report a throughput of 451 Mbps of their fast configuration and 115 Mbps for an economic configuration. A drawback of their design is the missing on-chip round-key generation. Chodowiec et al. [2] presented an implementation for low-end devices. Using only few resources they achieve a throughput ranging from 139 Mbps to 166 Mbps depending on the used FPGA device.

All implementations (except [6, 2]) use a considerable amount of hardware resources. For instance, [9] requires 138 block RAMs for 256-bit keys. This demands the use of expensive million-gate FPGA devices.

As shown above, most published hardware implementations focus on high throughput rates and do not provide a non-parameterizable design to support the complete AES standard. Furthermore, the high throughput implementations [3, 9] do not support the Cipher Block Chaining mode (CBC).

## 3.2    Architecture of the AES FPGA Implementation

This section describes the architecture of the AES co-processor. Starting with a swift overview, we will present details to highlight some innovative improvements that make it possible to present a resource-efficient AES co-processor suitable for low-end FPGA devices.

Basic components of the AES co-processor as shown in Fig. 3 are the AMBA APB interface, the data unit, the key unit, and the control unit. The key unit calculates the KeyExpansion function. All round keys are pre-calculated and stored in the key unit. Pre-calculated round keys allow fast en-/decryption of different data blocks for the same cipher key because no additional KeyExpansion is required. The data unit holds the State and performs all AES transformations: AddRoundKey, (Inv)SubBytes, (Inv)ShiftRows and (Inv)MixColumns. When encryption or decryption has completed the ciphertext (plaintext in case of decryption, respect.) is stored in the data unit. The control unit receives commands from the AMBA interface and generates control signals for all other modules. In addition to control round-key calculation, encryption and decryption, it also sequences data loading and unloading.

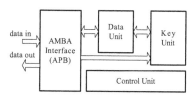

**Fig. 3.** Architecture of the AES co-processor

The architecture is similar to the architecture presented in Section 2. Differences are a modified State representation and a modified round-key calculation scheme. Due to a non-pipelined approach, the same performance for all modes of operations (ECB and CBC) is reached. Next we describe the AES data unit and the AES State representation in detail.

**Data Unit.** The data unit stores the State, all intermediate results of the round function applied to the State and the output data when encryption or decryption has completed. The major difference to all other published AES implementations is the innovative State representation that consists of two States. One State contains the actual State values and the other State stores newly calculated values. Figure 4 depicts the two States, referred to as StateA and StateB. In each cycle, 32 bits (one row or one column) of either StateA or StateB are altered. Using a second State provides a lot of benefits without the need of additional recourses: (Inv)ShiftRows comes for free and no State transposition between column and row operations is required.

Storage elements in FPGAs can be efficiently implemented by using synchronous RAMs because the basic logic elements of FPGAs, called slices, can

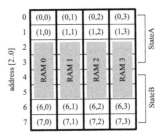

**Fig. 4.** The State-RAM

be configured as $16 \times 1$ bit synchronous RAM. Two slices (= one Configurable Logic Block - CLB) provide $16 \times 1$ bit synchronous dual-port RAM functionality (see [20]). Dual-port RAMs allow concurrent reading and writing to the RAM. Due to these technology features, the State-RAM as depicted in Fig. 4 is implemented as four slices of $8 \times 8$ bit synchronous dual-port RAMs to allow addressing the slices independently.

The data unit performs all transformations of the round function: (Inv)Shift-Rows, (Inv)SubBytes, (Inv)MixColumns and AddRoundKey. AddRoundKey and (Inv)MixColumns are applied to the State column by column, whereas (Inv)-ShiftRows and (Inv)SubBytes are applied to the State row by row. Due to the slice architecture of the RAM which holds the State, it is not possible to read/write from/to the RAM column by column. Hence, a transposition of the State is necessary if a row-oriented operation follows a column-oriented operation, or vice versa. Transposition would require a reorganization of the State before further operations can be performed. By using two States, transposition can be implemented by accordingly addressing the State-RAM. Furthermore, (Inv)ShiftRows can be combined with transposing the State. As a consequence of this, (Inv)ShiftRows and transposition come for free. In the sequel we describe the memory organization and State transposition for encryption. The same approach can easily be modified for decryption.

When a row-oriented operation follows a column-oriented operation (or vice versa), the State must be transposed. Combining row and column transformations minimizes the number of required transpositions: ShiftRows is combined with SubBytes and AddRoundKey is combined with MixColumns. This approach requires only one transposition per round. Encryption requires Sub-Bytes followed by ShiftRows. Since ShiftRows does not affect the byte values and SubBytes is applied to each byte of the State individually, the order of both operations does not matter. This fact eases the address generation for the State-RAM.

For explaining the State transposition we consider the State as $4 \times 4$ matrix: $\mathbf{S} = (s_{i,j})_{j=0..3}^{i=0..3}$. The ShiftRows transformation described in [10] can then be expressed as follows:

$$\mathbf{S}' = ShiftRows(\mathbf{S}) = (s_{i,j-i \bmod 4})_{j=0..3}^{i=0..3} \tag{1}$$

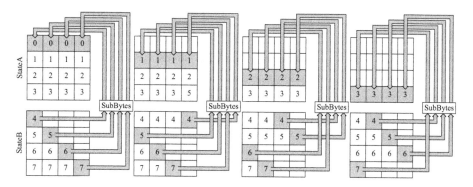

**Fig. 5.** ShiftRows and SubBytes for encryption

If we replace the State by the transposed State, we obtain:

$$\mathbf{S'^T} = ShiftRows(\mathbf{S^T}) = (s_{i+j\,mod\,4,j})_{j=0..3}^{i=0..3} \tag{2}$$

With the result of (2) the addressing of the StateB-RAM can be determined: the indices $(i,j)$ must be substituted with $(i+j\,mod\,4,j)$. Due to the even number of AES rounds for all key lengths, ShiftRows is always applied to StateB only. Thus, the resulting index tuples can be directly mapped to the RAMs. The first part of the tuple index specifies the RAM slice and the second part specifies the RAM address. Since we operate on StateB, we must add an offset of 4 to the index value to get the correct address. Figure 5 shows the transposition of the State, including ShiftRows and SubBytes for encryption.

**Implementation of (Inv)SubBytes and (Inv)MixColumns.** The (Inv)-SubBytes transformation is based on [18]. One difference is that the byte inversion in GF($2^8$) is implemented by using a synchronous ROM. (Inv)MixColumns is similar to the architecture presented in [17]. For further details refer to [18, 17].

**Key Unit.** An innovative aspect of our implementation is that the key unit can handle 128-bit, 192-bit and 256-bit keys with minimal additional hardware requirements. Supporting all key lengths increases the needed hardware resources for the key unit by only 7.8%. The size of the key memory for 256-bit keys is the same as for 128-bit keys. For 128-bit keys, the KeyExpansion function derives 44 32-bit round-key parts from the cipher-key. This requires a 64 × 32 bit RAM. 256-bit keys produce 63 32-bit round-key parts fitting the 64 × 32 bit RAM.

### 3.3    Performance of the FPGA AES Implementation

This section compares the proposed AES co-processor with the works referred to in Section 3.1. In order to provide comparable results, we implemented our co-processor on a Xilinx Virtex-E XCV1000EBG560-8 device.

**Table 4.** Hardware resources and throughput comparison

| Work | Device | #CLB-slices | #BRAM | ECB mode Throughput [Mbps] |
|---|---|---|---|---|
| Gaj et al. [3] | Xilinx XCV1000 | 12,600 | 80 | 12,160 |
| McLoone et al. [9] (I) | Xilinx XCV812E | 2,222 | 100 | 6,956 |
| McLoone et al. [9] (II) | Xilinx XCV3200E | 2,577 | 112 | 5,800 |
| McLoone et al. [9] (III) | Xilinx XCV3200E | 2,995 | 138 | 5,000 |
| McLoone et al. [9] (IV) | Xilinx XCV3200E | 7,576 | 102 | 3,239 |
| Dandalis et al. [5] | Xilinx XCV1000 | 5,673 | ? | 353 |
| Fischer et al. [6] (I) | FLEX 10KE200-1 | 2,530 | 24 | 451 |
| Fischer et al. [6] (II) | ACEX 1K50-1 | 1,213 | 10 | 115 |
| Chodowiec et al. [2] | Xilinx XC2S30-6 | 222 | 3 | 166 |
| **Our proposal** | Xilinx XCV1000E | **1,125** | **0** | **215** |

[9]: enc.: (I)AES-128, (II)AES-192, (III)AES-256, enc./dec.:(IV)AES-128
[6]: AES-128 enc./dec.: (I) fast configuration, (II) economic configuration

The performance results given in Table 4 are for the ECB mode. Most of these implementations claiming high throughput rates will have similar performance figures when operating in the CBC mode. The CBC mode is strictly recommended and commonly used for encrypting high-speed data streams (*e.g.* as it is used for encrypting data transfers over networks) and hence, the above-listed high throughput rates lose their significance.

As shown in Table 4 our implementation is the only one that does not require any block RAMs. The presented AES co-processor supports the complete AES standard and the CBC mode. Additionally, it is equipped with a 32-bit AMBA APB interface that eases the integration with processors used in System-on-Chip designs [1]. If we do not consider the CBC mode and the AMBA bus interface, our approach is still comparable with the above-listed works but we would require less hardware resources (-26 %).

Our implementation utilizes 9.16% of the available logic cells on a Xilinx Virtex-E XCV1000EBG560-8 device. 90.8% of the logic resources and 100% of the on-chip BRAMs can be used by other circuits like a LEON2 or an ARM processor. For a stand-alone application a low-end FPGA (*e.g.* Xilinx SpartanII XC2S100-6) is sufficient for implementing the complete AES co-processor— the other approaches (except [2]) do not fit on a SpartanII device. The high throughput designs do not support this flexibility and require expensive multi-million gate FPGAs. Another important fact is that the other works do not provide an en-/decryption engine that supports all defined key lengths.

The maximum clock frequency on a XCV1000 FPGA is 161 MHz. At this frequency, a throughput of 215 Mbps for AES-128, 180 Mbps for AES-192, and 156 Mbps for AES-256 is achieved for both ECB mode and CBC mode.

# 4    Conclusions

In this article we presented two designs of a compact AES co-processor, one suitable for ASIC implementations, the other one suitable for FPGAs. Both designs are able to implement the whole functionality of the AES standard: encryption and decryption with all key lengths (128-bit, 192-bit, and 256-bit). In addition to covering the complete AES standard they support the Cipher Block Chaining mode CBC. The AES co-processors also have a standard 32-bit interface (AMBA) that facilitates the integration in System-on-Chip designs.

Our ASIC implementation is very regular, which makes it well suited for full custom designs, and highly scalable. By scaling, the ASIC AES module it can be adapted for many different applications with different requirements. With this architecture high performance (up to 198 Mbps) as well as low area requirements (down to 8,500 GE) can be reached on a 0.6 $\mu m$ technology.

For the FPGA implementation, we have shown that due to an innovative State representation the complete AES co-processor can be implemented on inexpensive low-end FPGA devices. An implementation on a Xilinx Virtex-E device uses only 1,125 CLB-slices and no block RAMs. Our FPGA implementation reaches a throughput of 215 Mbps at a clock frequency of 161 MHz for encryption and decryption.

# References

1. ARM Limited. AMBA 2.0 Specification. "http://www.arm.com/armtech/", 2001.
2. P. Chodowiec and K. Gaj. Very Compact FPGA Implementation of the AES Algorithm. In *Workshop on Cryptographic Hardware and Embedded Systems – CHES 2003*, volume 2779 of *Lecture Notes in Computer Science (LNCS)*, pages 319–333. Springer, 2003.
3. P. Chodowiec, P. Khuon, and K. Gaj. Fast Implementations of Secret-Key Block Ciphers Using Mixed Inner- and Outer-Round Pipelining. In *Symposium on Field Programmable Gate Arrays – FPGA 2001*, pages 94–102. ACM Press, 2001.
4. J. Daemen and V. Rijmen. *The Design of Rijndael.* Springer-Verlag, 2002.
5. A. Dandalis, V. Prasanna, and J. Rolim. A Comparative Study of Performance of AES Final Candidates Using FGPAs. "http://csrc.nist.gov/CryptoToolkit/aes/round2/conf3/aes3agenda.html", 2000.
6. V. Fischer and M. Drutarovský. Two Methods of Rijndael Implementation in Reconfigurable Hardware. In *Workshop on Cryptographic Hardware and Embedded Systems – CHES 2001*, volume 2162 of *Lecture Notes in Computer Science (LNCS)*, pages 77–92. Springer-Verlag, 2001.
7. P.C. Kocher, J. Jaffe, and B. Jun. Differential Power Analysis. In *Advances in Cryptology – CRYPTO 1999*, volume 1666 of *Lecture Notes in Computer Science*, pages 388–397. Springer-Verlag, 1999.
8. S. Mangard, M. Aigner, and S. Dominikus. A Highly Regular and Scalable AES Hardware Architecture. In *IEEE Transactions on Computers*, volume 52, pages 483–491, April 2003.

9. M. McLoone and J.V. McCanny. High Performance Single-Chip FPGA Rijndael Algorithm Implementations. In *Workshop on Cryptographic Hardware and Embedded Systems – CHES 2001*, volume 2162 of *Lecture Notes in Computer Science (LNCS)*, pages 65–76. Springer-Verlag, 2001.

10. National Institute of Standards and Technology. Federal Information Processing Standard 197, The Advanced Encryption Standard (AES). "http://csrc.nist.gov/publications/fips/fips197/fips-197.pdf", 2001.

11. N. Pramstaller, F.K. Gürkaynak, S. Haene, H. Kaeslin, N. Felber, and W. Fichtner. Towards an AES Crypto-chip Resistant to Differential Power Analysis. In *Proccedings of ESSCIRC 2004, to appear*, 2004.

12. A. Rudra, P.K. Dubey, C.S. Jutla, V. Kumar amd J.R. Rao, and Pankaj Rohatgi. Efficient Rijndael Encryption Implementation with Composite Field Arithmetic. In *Workshop on Cryptographic Hardware and Embedded Systems – CHES 2001*, volume 2162 of *Lecture Notes in Computer Science (LNCS)*, pages 171–184. Springer-Verlag, 2001.

13. A. Satoh, S. Morioka, K. Takano, and S. Munetoh. A Compact Rijndael Hardware Architecture with S-Box Optimization. In *Advances in Cryptology – ASIACRYPT 2001*, volume 2248 of *Lecture Notes in Computer Science (LNCS)*, pages 239–254. Springer-Verlag, 2001.

14. K. Tiri, M. Akmal, and I. Verbauwhede. A Dynamic and Differential CMOS Logic with Signal Independent Power Consumption to Withstand Differential Power Analysis on Smart Cards. In *Proceedings of ESSCIRC 2002*, 2002.

15. K. Tiri and I. Verbauwhede. Securing Encryption Algorithms against DPA at the Logic Level: Next Generation Smart Card Technology. In *Workshop on Cryptographic Hardware and Embedded Systems – CHES 2003*, volume 2779 of *Lecture Notes in Computer Science (LNCS)*, pages 125–136. Springer, 2003.

16. B. Weeks, M. Bean, T. Rozylowicz, and C. Ficke. Hardware Performance Simulations of Round 2 Advanced Encryption Standard Algorithms. "http://csrc.nist.gov/encryption/aes/round2/NSA-AESfinalreport.pdf", 2000.

17. J. Wolkerstorfer. An ASIC implementation of the AES-MixColumn operation. In *Proceedings of Austrochip 2001*, October 2001.

18. J. Wolkerstorfer, E. Oswald, and M. Lamberger. An ASIC implementation of the AES S-Boxes. In *Topics in Cryptology – CT-RSA 2002, Proceedings of the RSA Conference 2002*, volume 1965 of *Lecture Notes in Computer Science*. Springer-Verlag, February 2002.

19. S-Y Wu, S-C Lu, and C-S Laih. Design of AES Based on Dual Cipher and Composite Field. In *Topics in Cryptology – CT-RSA 2004, Proceedings of the RSA Conference 2004*, volume 2964 of *Lecture Notes in Computer Science*. Springer-Verlag, February 2004.

20. Xilinx Incorporated. Silicon Solutions — Virtex Series FPGAs. "http://www.xilinx.com/products/".

# Small Size, Low Power, Side Channel-Immune AES Coprocessor: Design and Synthesis Results

Elena Trichina[1], Tymur Korkishko[2], and Kyung Hee Lee[2]

[1] Department of Computer Science, University of Kuopio,
P.O.B. 1627, FIN-70211, Kuopio, Finland**
[2] Information security TG, i-Networking Lab, Information Security Group,
Samsung Advanced Institute of Technology, Korea

**Abstract.** When cryptosystems are being used in real life, hardware and software implementations themselves present a fruitful field for attacks. Side channel attacks exploit information such as time measurements, power consumption, and electromagnetic emission that leaks from a device when it executes cryptographic applications. When leaked information is correlated to a secret key, an adversary may be able to recover the key by monitoring this information. This paper describes an AES coprocessor that provides complete protection against first-order differential power analysis by embedding a widely used software countermeasure that decorrelates data being processed from the leaked information, so-called data masking, at a hardware level.

## 1   Introduction

In applications such as smart cards hardware complexity and tamper resistance are very important issues that directly affect the cost and consumer acceptance of such devices. A class of *side channel* attacks enables breaking cryptographic algorithms by measuring timing characteristics [12], power consumption [11, 18], and electromagnetic radiation [8, 23] of a smart card microprocessor when it runs cryptographic applications.

Until recently, most of these attacks exploited some specific features of software implementations of cryptographic algorithms, and many countermeasures were designed at a software level. For many applications, however, it is necessary that cryptographic algorithms should be realized in hardware. Although not many results have been published yet, it is prudent to suggest that cryptographic hardware also leaks side channel information, and that alongside with general tamper-resistant features cryptographic coprocessors should include countermeasures specifically designed to protect them against side channel attacks.

One of the most powerful software techniques to counteract such attacks is to *mask* all input and intermediate data values in order to de-correlate any information leaked through side channels from actual secret data being processed.

---

** This work had been done when the author was with the Smart Card System Engineering Business Unit, System LSI Division, Samsung Electronics Co. LTD., Korea.

H. Dobbertin, V. Rijmen, A. Sowa (Eds.): AES 2004, LNCS 3373, pp. 113–127, 2005.

In [28] it had been shown how to apply data masking technique at the level of micro operations such as logical AND, XOR, etc., and how to use these operations as building blocks for implementation of inversion in composite fields directly on masked data. In this paper we use ideas from [28] in order to build a practical coprocessor for an Advanced Encryption Standard algorithm [7] that is immune to side channel attacks. First, we generalize multiplication on masked data from a bitwise multiplication operation (i.e., logical AND) to multiplication in any $GF(2^n)$ field. This generalization allows us to use standard libraries while designing side channel attacks-resistant cryptographic hardware for applications that are based on binary field arithmetic. What is more important, our method provides considerable savings in the gate count and power consumption in comparison with [28].

The synthesis results for a "minimalist" architecture that implements a 16 clock/round version of the AES coprocessor with a flexible key size, show that with 0.18 $\mu m$ technology the performance of 4Mbps can be achieved with the circuit comprising 20,506 gates clocked at 5MHz and requiring 1.07 $\mu A$. This is better than countermeasures such as dynamic and differential CMOS logic [26] or asynchronous circuits with dual rail logic [21] can offer. Also, our solution has an advantage of using standard technologies, standard libraries, and well-established design tools.

The rest of the paper is organized as follows. After a brief reminder of the AES algorithm in Chapter 2, we describe principles of power analysis attacks. The countermeasure consisting in masking all input and intermediate data with some random values and difficulties of its implementation for the AES algorithm are discussed in Chapter 4, after which we suggest our solution to this problem. The details of the DPA-resistant AES coprocessor architecture are given in Chapter 6. The paper is concluded with synthesis results and a brief comparison of our design with other hardware solutions.

## 2    AES Reminder

The Advanced Encryption Standard [7] is a round-based symmetric block cipher. The standard key size is 128 bits, but for some some applications 192 and 256-bit keys must be supported as well. The round consists of four different operations, namely, SubBytes, ShiftRows, MixColumn, and AddRoundKey, that are performed repeatedly in a certain sequence; each operation in a standard algorithm maps a 128-bit input state into a 128-bit output state. The state is represented as a 4 × 4 matrix of bytes. The number of rounds depends on the key size. In the decryption process, the inverse operations are executed in a slightly different order.

ShiftRows is a cyclic shift operation on each of four rows in a 4 × 4-byte state using 0 ∼ 3 offsets. MixColumn treats 4-byte data blocks in each column of a state as coefficients of a 4-term polynomial, and multiplies them modulo $x^4 + 1$ with the fixed polynomial $c(x) = \{03\}x^3 + \{01\}x^2 + \{01\}x + \{02\}$. AddRoundKey is a bit-wise XOR operation on 128-bit round keys and data. These three operations are linear.

*SubBytes* is the main building block of the AES. It replaces each byte in a state by its substitute in an *Sbox* that comprises a composition of two transformations:

- First, each byte in a state is replaced with its reciprocal in the finite field $GF(2^8)$, except that 0, which has no reciprocal, is replaced by itself. This is the only *non-linear* function in the AES algorithm.
- Then an affine transformation $f$ is applied. It consists of a bitwise matrix multiply with some fixed $8 \times 8$ binary matrix followed by XOR with the hexadecimal number {63}.

The *round key* is computed in parallel to the round operation. It is derived from the cipher key by means of key expansion and round key selection operations, which are similar to those of the round operations, and also use *Sboxes*.

## 3   S-Box Architecture

There are many design trade-offs to be considered when implementing an S-box in ASIC since the size, the speed, and the power consumption of an AES coprocessor depends largely on the number and the style of implementation of *Sboxes* [20]. It also turned out that this operation is the most difficult to protect against side channel attacks.

To optimize the silicon area, a number of flexible ASIC solutions that use similarities between encryption and decryption to share silicon were proposed [14, 20], where *SubBytes* was implemented in two steps, as a combination of inversion in the field and an affine transformation $f$. While the affine transformations used for encryption and decryption are slightly different, the silicon implementing inversion in $GF(2^8)$ can be used for both, as shown in Fig. 1. Therefore, an area- and power-efficient and secure implementation of inversion may have a big impact on an overall design.

The most obvious solution is to use a look-up table for this operation [14]. It is fast and inexpensive in terms of power consumption [20]. There is a major drawback, however. Namely, the size of silicon is about 1,700 gate equivalents per table in $0.18\mu$ technology. Considering that up to 20 such tables (including 4 tables for key scheduling) is required per round, this solution is hardly feasible for co-processors intended for smart cards and other embedded systems.

Among various alternative approaches composite field inversion produces the most compact AES implementations [24, 25, 30] which can be further optimized

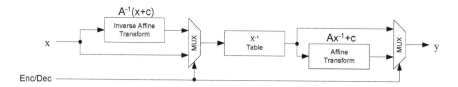

**Fig. 1.** S-box implementation suitable for encryption and decryption

to minimize power consumption [20]. As a basis for our design we use fully combinational logic implementation of inversion in composite fields described in [30].

Usually the field $GF(2^8)$ is seen as an extension of $GF(2)$ and therefore its elements can be represented as bytes. However, $GF(2^8)$ can also be seen as a quadratic extension of $GF(2^4)$; in this case an element $a \in GF(2^8)$ is represented as a linear polynomial $a_H x + a_L$, denoted $[a_H, a_L]$, with coefficients in $GF(2^4)$. This isomorphic representation is far better suited for hardware implementation [22, 24, 20, 30].

The bijection from $a \in GF(2^8)$ to a two-term polynomial $[a_H, a_L]$ is given by the linear function $map$ computed by means of XOR operations on bits of $a$. The inverse transformation $map^{-1}$ converts a two-term polynomial back to element $a \in GF(2^8)$, and is defined in a similar way. For more details see [30].

All arithmetic operations applied to elements of $GF(2^8)$ can also be computed in a new representation. Two-term polynomials are added by addition of corresponding coefficients. Multiplication and inversion of a two term-polynomial requires modular reduction to ensure that the result is a two-term polynomial as well; the irreducible polynomial $n(x) = x^2 + \{1\}x + \{e\}$ whose coefficients are chosen to optimize finite field arithmetic can be used for this purpose.

Inversion of a two-term polynomial is defined as $(a_H x + a_L) \otimes (a_H x + a_L)^{-1} = \{0\}c + \{1\}$. From this definition the formulae for inversion can be derived:

$$(a_H x + a_L)^{-1} = (a_H \otimes d)x + (a_H \oplus a_L) \otimes d \qquad (1)$$
$$d = ((a_H^2 \otimes \{e\}) \oplus (a_H \otimes a_L) \oplus a_L^2)^{-1}.$$

Fig. 2 depicts a block diagram of inversion in composite field $GF((2^4)^2)$. As one can see, one addition, one squaring, one multiplication by a constant, three general multiplications, and one inversion in $GF(2^4)$ are necessary for such implementation. All these operations can be realized in combinational logic. Only general multiplication and inversion in $GF(2^4)$ require to use both, AND and XOR gates for their implementation; all other operations need only compositions of XOR gates [30].

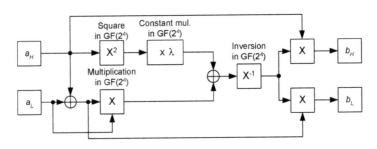

**Fig. 2.** Inversion in $GF((2^4)^2)$

## 4   Side Channel Attacks and Computations on Masked Data

Basically, side-channel attacks work because there is a correlation between the physical measurements taken during computations, such as power consumption [11, 18], EMF radiation [8, 23], time of computations [12], and the internal state of the processing device, which itself is related to a secret key.

### 4.1   Power Analysis Basics

Among side-channel attacks, a *differential power analysis* (DPA) is the main concern when implementing cryptographic algorithms in embedded devices because due to physical constraints an adequate shielding and power consumption filtering cannot be employed. Power analysis attacks use an all-pervasive fact that, ultimately, all calculations performed by a digital device operate on logical ones and zeros; and in contemporary technology power consumption while manipulating a logical one differs from power consumption while manipulating a logical zero.

To illustrate why the power analysis works, let us consider an example in Fig. 3. The circuit, first appeared in [18], represents a component model useful for understanding power consumption characteristics of Complementary Metal Oxide Semiconductor (CMOS) technology.

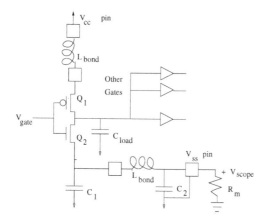

**Fig. 3.** Measuring power consumption of a smart card

The two most essential components of power consumption during the change of a state of a CMOS gate are dynamic charge resp. discharge (appr. 85%) and dynamic short circuit current (appr. 15%). This is sketched in Fig. 3 where the output of each gate has a capacitive load consisting of the parasitic capacity of the connected wires and gates of the following stages. An input transition results in an output transition, which discharges or charges this parasitic capacity, causing a current to flow to $V_{cc}$ or to $V_{ss}$.

For example, as $V_{gate}$ changes from 0 to 5 volts, the transistors $Q_1$ and $Q_2$ are both conducting for a brief period causing current to flow from $V_{cc}$ to the ground. Also, during this time the capacitor $C_{load}$ will be discharged (or charged) causing more (or less) current to flow through the $V_{ss}$ pin. The current charges and discharges capacitors $C_{load}$, $C_1$, and $C_2$, and flows out of the smart card through a bond wire that acts as an inductor $L_{bond}$.

Power dissipated by the circuit can be monitored by using a small resistor $R_m$ in series between the $V_{ss}$ pin and the ground (or, alternatively, between the $V_{cc}$ pin and the true source). Current moving through $R_m$ creates a time varying voltage that can be sampled by a digital oscilloscope.

The more circuit changes its state, the more power is dissipated. In a synchronous design gates are clocked, which means that all gates change their state at the same time. Information useful to an attacker is leaked because the amount of current being drawn when the circuit is clocked is directly related to a change of the state of $C_{load}$ or the current drawn by other gates attached to $C_{load}$. Thus, whenever the secret key data or data correlated to the secret key is manipulated, the microprocessor can leak damaging information that can be observed at the $V_{scope}$.

In a microprocessor each clock pulse causes many bit transitions to occur simultaneously. There are two types of information leakage that can be observed at the $V_{scope}$: Hamming weight leakage and transition count leakage [18]. The Hamming weight information leaks when the dominant source of current is caused by discharging of the $C_{load}$. The transition current information can leak when the dominant source of current is due to the switching of the gates that are driven by the data bus. When the data bus changes state, many of the gates driven by the bus will briefly conduct current. Thus, the more bits change states, the more power is dissipated.

From this explanation we deduce that the power consumption of a circuit at time $t$ is the sum of power dissipated by all gates at this time. Of course, various noise components must be considered as well [18]. It can be stated as the simple power model

$$P(t) = \sum f(g, t) + N(t),$$

where $t$ denotes time, $N(t)$ is a normally distributed random variable which represents a noise component, and $f(g, t)$ denotes power consumption of gate $g$ at time $t$.

The next step is to relate this model to statistics. If we consider function $f(g, t)$ as a random variable from an unknown probability distribution, then according to the Central Limit Theorem, $P(t)$ is normally distributed. In a DPA attack, an attacker divides the power measurements in two or more different sets and tries to compute the difference between these sets in order to verify the *selection function*, which relates the corresponding power measurements to the hypothesis concerning the values of the target bits of the key. Only if the hypothesis was correct, there will be some noticeable peaks in statistics.

For example, a selected bit $b$ at the output of one *Sbox* of the first round of the AES algorithm will depend on the known input message and 8 unknown bits of the key. The correlation between power consumption and $b$ can be computed for all 256 values of 8 unknown bits of the key. The correlation is likely to be maximal for the correct guess of the 8 bits of the key. Then an attack can be repeated for the remaining *Sboxes*.

## 4.2  Data Masking and Inversion in $GF(2^4)$

There are many strategies to combat side-channel attacks. On a hardware level, the countermeasures usually include clock randomization, power consumption randomization, current compensation, and various detectors of abnormal behavior. However, the effect of these countermeasures can be reduced by various signal processing and statistical techniques [6]. Software-based countermeasures include introducing dummy instructions, randomization of the instruction execution sequence, balancing Hamming weights of the internal data, etc.

*Data masking* is one of the most powerful software countermeasures against side channel attacks [5, 11]. The idea is simple: the message and the key are masked with some random values at the beginning of computations, and thereafter everything is almost as usual. Of course, the value of the mask at the end of some fixed step (e.g., at the end of a round or at the end of a linear part of computations) must be known in order to re-establish the expected data value at the end of the execution; we call this *mask correction*.

A traditional XOR operation is used for data masking; however, a mask is arithmetic in $GF(2^8)$. The operation is compatible with the AES structure except for inversion in *SubBytes*, which is the only non-linear transformation. In other words, to compute mask corrections for each of the linear transformations in a round, we simply have to apply this transformation separately to masked bytes and to corresponding masks.

Unfortunately, it turned out to be rather difficult to find an efficient and secure solution for non-linear operations. The first attempt to transform masked data between Boolean and arithmetic operations in a secure way [17] was shown

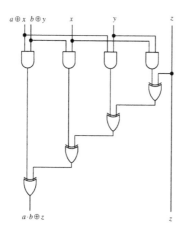

**Fig. 4.** Masked AND (MAND)

to be insufficient against DPA attacks. A more sound method for mask switching [10] involves too much computational overhead.

To overcome this difficulty, Akkar and Giraud [1] proposed transformed masking, where first an additive mask is replaced by a multiplicative mask in a series of multiply and add operations, after which a normal inversion takes place, and finally, the transformation of a multiplicative mask into an additive mask is carried out again. However, it was pointed out in [9] that a multiplicative mask does not blind zero, and thus does not prevent a DPA attack.

In [28] it was noticed that for a fully combinational AES *Sbox* design, the problem of "masked inversion" can be effectively reduced to the problem of computing binary XOR and AND operations on masked bits $\tilde{a} = a \oplus x$, $\tilde{b} = b \oplus y$ and on bits of the mask $x$, $y$ without ever revealing actual data bits $a$, $b$ in the process. For XOR computing the mask correction is trivial because $\tilde{a} \oplus \tilde{b} = (a \oplus b) \oplus (x \oplus y)$.

Masked AND and the corresponding mask correction can be computed by manipulating only **masked data** bits and the bits of the masks as follows [1]. Let $c = a \cdot b$. Then

$$c \oplus z = a \cdot b \oplus z = (\tilde{a} \cdot \tilde{b} \oplus (y \cdot \tilde{a} \oplus (x \cdot \tilde{b} \oplus (x \cdot y \oplus z)))). \qquad (2)$$

This can be implemented as a cascade of logic gates as shown in Fig. 4.

To realize inversion in $GF(2^8)$ on masked data, one would have to replace each AND gate used by a multiplier and an inverter in $GF(2^4)$ with the circuit depicted in Fig.4, increasing more than fourfold a total amount of gates in the *Sbox* combinational logic design.

## 4.3    Masked $GF(2^n)$ Multiplier

Let us make an observation that equation (2) can be generalized to the equation for "masked multiplication' in any field $GF(2^n)$. If $A, B, X, Y, Z \in GF(2^n)$, then

$$(A \otimes B) \oplus Z = [(A \oplus X) \otimes (B \oplus Y)] \oplus [X \otimes (B \oplus Y)] \oplus [(A \oplus X) \otimes Y] \oplus [X \otimes Y] \oplus Z. \quad (3)$$

Hence, masked multiplication in $GF(2^n)$ can be performed using *conventional multipliers* with any architecture, and four additional XOR operations for mask correction, as depicted in Fig. 5. An additional mask $Z$ is also used to mask the output product. Hence, this approach requires 4 "normal" multipliers in $GF(2^n)$ and 4 bitwise XOR operations.

In contrast, a straightforward application of the technique suggested in [28] involves building masked multipliers by replacing every AND gate in a "normal" multiplier with a masked AND( MAND) circuit. For example, Fig.6 illustrates a transformation of a conventional multiplier in $GF(2^2)$ into a masked multiplier using this approach.

---

[1] To achieve balanced and independent intermediate results, the scheme is used a freshly generated random bit $z$ as a new mask.

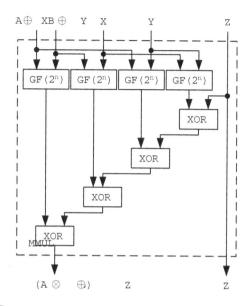

**Fig. 5.** Building a masked multiplier in $GF(2^n)$ from standard multipliers

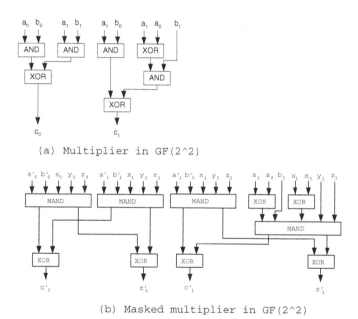

**Fig. 6.** Transformation of a usual (a) multiplier into a masked (b) multiplier in $GF(2^2)$

The implementation complexity of a masked multiplier can be calculated knowing the complexity of a basic standard multiplier. For example, for the popular Mastrovito multipliers in $GF(2^n)$ [16], the space complexity expressed in the number of AND and XOR gates is $n^2$ and $\geq (n^2 - 1)$, respectively. Then the complexity of the generalized masked multiplier when compared with the original straightforward implementation constitutes 9 to 29 % improvement in the number of XOR gates, as can be seen from the Table 1.

**Table 1.** Comparison of the space complexity of generalized and straightforward masked multipliers in $GF(2^n)$

| GF n | Irreducible Polynomial | Mastrovito mult. AND | XOR | Straight. masked AND | XOR | Proposed mult. AND | XOR | Advantage % |
|------|------------------------|---------------------|-----|----------------------|-----|--------------------|-----|-------------|
| 2  | [2,1,0]       | 4   | 3   | 16   | 22   | 16   | 20   | 9.1  |
| 4  | [4,1,0]       | 16  | 15  | 64   | 94   | 64   | 76   | 19.1 |
| 8  | [8,5,3,2,0]   | 64  | 84  | 256  | 382  | 256  | 284  | 25.7 |
| 16 | [16,11,6,5,0] | 256 | 281 | 1024 | 1534 | 1024 | 1084 | 29.3 |

## 5   Secure AES Coprocessor

When manipulating masked data, all operations in a round, apart from *Sub-Bytes*, require simple mask corrections in a form of analogous computations on masks that are carried out in parallel with the main computation flow. This can be achieved simply by duplicating hardware for all transformations but *Sub-Bytes*.

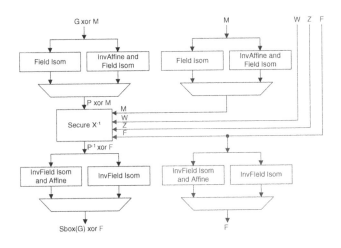

**Fig. 7.** Masked S-box

Another solution is to pipeline computations on masked data and on masks, which halves the throughput, and for which additional 128-bit registers are required for mask values. Since each 1-bit register needs an equivalent of 6-7 gates, there is no visible advantage for pipelining.

The structure of the *Masked Sbox* is depicted in Fig. 7. The inverse field isomorphism $map^{-1}$ and affine transformation $f$ are both linear operations, and they are merged to optimize the gate count; the same holds for $map$ and $f^{-1}$ in decryption [14, 20]. A duplicate data path on the right hand side computes a mask correction.

In this figure, the box "Secure $X^{-1}$" represents inversion in $GF((2^4)^2)$ implemented as was defined in Eq. 1. $M$ is a random mask, $(G \text{ xor } M)$ represents a masked input, $Z$, $W$, $F$ are new masks used to "refresh" the masked data at the end of each operation in $GF(2^4)$ and at the end of inversion in $GF((2^4)^2)$. The details are given in Fig. 8 where every SMul$2^n$ box is, in fact, a generalized masked multiplier in $GF(2^4)$. The box $X^{-1}$ represents inversion in $GF(2^4)$, and is implemented in combinational logic in the same way as in [30], with every AND being replaced with masked AND.

Altogether, 1.2K gate equivalent is required to implement one S-box which is 25% better than the table lookup implementation.

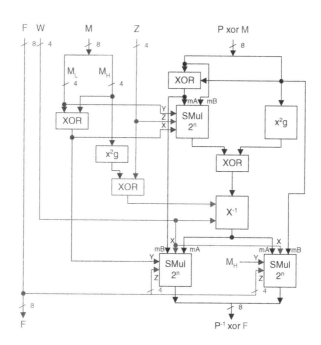

**Fig. 8.** Masked inverter in $GF((2^4)^2)$

## 6    Conclusion

The secure hardware AES module based on the described scheme has been fully implemented in 0.18$\mu m$ technology. As a balance between the throughput and the gate count, 1 *Sbox* has been used for the main datapath, and one for the key scheduling. Our implementation of *MixColumn* resembles the one reported in [29] and exploits common subexpressions for MixColumn/InvMixColumn operations.

The general design flow for a secure AES module is depicted in Fig. 9. First, a Verilog model had been created and tested, after which Cadence Design System Verilog-XL simulator was used to generate timing diagrams and to verify the correctness of the design. When RTL code had been verified, the digital circuit was synthesized with Synopsys Design Analyzer tool using Samsung 0.18 $\mu m$ libraries. Power-compiler simulation data at 5 MHz were obtained with simulation tool CubicWare.

The summary of the synthesis results are given in the table in Fig. 10. The total gate count for the secure AES module (excluding I/O) with a flexible key size is 20,06K, while for a standard 128-bit key the gate count can be reduced to 16K. The power consumption is 2.0 $mA$ in 0.18$\mu m$ technology for a flexible key size architecture, and 1.6 $mA$ for a 128-bit key standard. With 0.13$\mu m$ technology, the power consumption can be reduced to 1.1 $mA$, which allows us to use this module in applications such as GSM and ad-hoc networks. The secure AES module has throughput 4 Mbps when operated at 5 MHz.

The described approach can be applied to other cryptographic coprocessors that use arithmetic operations in Galois fields. It provides comparable protection as dynamic and differential logic [26] and asynchronous dual rail circuits [21] at the similar price in terms of the gate count and power consumption. Taking into

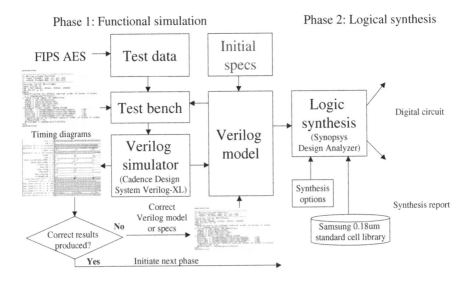

**Fig. 9.** Hardware design flow

| Components | Subcomponents | Scalable key size | 128 bit key size |
|---|---|---|---|
| Masked Sbox | Inverter in GF((2⁴)²) | 1206 | 1206 |
| | Input-output processing | 350 (180) | 350 (180) |
| Subtotal for masked Sbox (for key scheduler) | | **1556 (1386)** | **1556 (1386)** |
| Masked datapath | Masked data | 4900 | 4900 |
| | Mask | 2732 | 2732 |
| Subtotal for masked datapath | | **7632** | **7632** |
| Masked key scheduler | Masked key | 6523 | 4835 |
| | Mask | 4913 | 3225 |
| Subtotal for masked key scheduler | | **11436** | **8060** |
| Control | Data control | 300 | 300 |
| | Key control | 1140 | 400 |
| **TOTAL for masked AES** | | **20506** | **16390** |

**Fig. 10.** Gate count for a secure AES module

account that the latter techniques require new logic libraries, careful "balancing" of place and routing and new development tools which implies higher design and production costs and longer time-to-market, our solution offers a competitive alternative to a hardware protection.

# References

1. Akkar, M., Giraud, C.: An implementation of DES and AES, secure against some attacks. Proc. Cryptographic Hardware and Embedded Systems: CHES 2001. Lecture Notes in Computer Science **2162** (2001) 309-318
2. Anderson, R., Kuhn, M.: Low cost attacks on tamper resistant devices. Proc.Security Protocols: IWSP 1997. Lecture Notes in Computer Science **1361** (1997) 125-136
3. Blömmer, J., Merchan J. G., Krummel, V.: Provably secure masking of AES. IACR Cryptology ePrint Archive Report 2004/101 (2004)
4. M. Bucci, L.Germani, M. Guglielmo, R. Luzzi, A. Trifiletti: A simulation methodology for DPA resistance testing of cryptographic processors (manuscript) 2003
5. Chari, S., Jutla, C., Rao, J., Rohatgi, P.: Towards sound approaches to counteract power-analysis attacks. Proc. Advances in Cryptology – Crypto'99. Lecture Notes in Computer Science **1666** (1999) 398-412,
6. Clavier, C., Coron, J-S., Dabbous, N.: Differential power analysis in the presence of hardware countermeasures. Proc. Cryptographic Hardware and Embedded Systems: CHES 2000. Lecture Notes in Computer Science **1965** (2000) 252-263
7. Daemen, J., Rijmen, V.: *The design of Rijndael: AES - The Advanced Encryption Standard*. Springer-Verlag Berlin Heidelberg (2002)
8. Gandolfi, K., Mourtel, C., Oliver, F.: Electromagnetic analysis: concrete results. Proc. Cryptographic Hardware and Embedded Systems: CHES 2001. Lecture Notes in Computer Science **2162** (2001) 251-261
9. Golič, J., Tymen,Ch.: Multiplicative masking and power analysis of AES. Proc. Cryptographic Hardware and Embedded Systems: CHES 2002. Lecture Notes in Computer Science **2523** 198-212

10. Goubin, L.: A sound method for switching between boolean and arithmetic masking. Proc. Cryptographic Hardware and Embedded Systems: CHES'01. Lecture Notes in Computer Science **2162** (2001) 3-15
11. Kocher, P., Jaffe, J., Jun, B.: Differential power analysis. Proc. Advances in Cryptology – CRYPTO'99. K Lecture Notes in Computer Science **1666** (1999) 388-397
12. Kocher, P.: Timing attacks on implementations of Diffie-Hellmann, RSA, DSS, and other systems. Proc. Advances in Cryptology – Crypto'96. Lecture Notes in Computer Science **1109** (1996) 104-113
13. Kommerling, O., Kuhn, M.: Design principles for tamper-resistant smartcard processors. Proc. USENIX Workshop on Smartcard Technology (Smartcard 99) (1998) 9-20
14. Lu, C. C., Tseng, S-Y.: Integrated design of AES (Advanced Encryption Srandard) encryptor and decryptor. Proc. IEEE conf. on Application-Specific Systems, Architectures, and Processors (ASAP'02) (2002) 277-285
15. Mangard, S., Aigner, M., Dominikus, S.: A highly regular and scalable AES hardware architecture. IEEE Transactions on Computers **52** no. 4 (2003) 483-491
16. E.D. Mastrovito, *VLSI architectures for computations in Galois fields*, PhD Thesis, Linkoping University, Linkoping, Sweden (1991)
17. Messerges, T.: Securing the AES finalists against power analysis attacks. Proc. Fast Software Encryption Workshop 2000. Lecture Notes in Computer Science **1978** (2000) 150-165
18. Messerges, T. S., Dabbish, E. A., Sloan, R. H.: Examining smart-card security under the thread of power analysis. IEEE Trans. Computers. **51** no. 5 (2002) 541-522
19. Messerges, T. S.: Using second-order power analysis to attack DPA resistant software. Proc. Cryptographic Hardware and Embedded Systems – CHES 2000. Lecture Notes in Computer Science **1965** (2000) 238-251
20. Morioka, S., Satoh, A.: An optimized S-Box circuit architecture for low power AES design. Proc. Cryptographic Hardware and Embedded Systems: CHES 2002. Lecture Notes in Computer Science **2523** (2003) 272-186
21. Moore, S., Anderson, R., Cunningham, P., Mullins, R., Taylor, G.,: Improving smart card security using self-timed circuits. Proc. Proceeding 8th IEEE International Symposium on Asynchronous Circuits and Systems – ASYNC'02. IEEE (2002) 23-58
22. Paar, C.: *Efficient VLSI architectures for bit parallel computations in Galois fields*. PhD Thesis, University of Essen, Germany (1994)
23. Quisquater, J. J., Samide, D.: Electromagnetic analysis (ema): measures and counter-measures for smart cards. Proc. Smartcard Programming and Security. Lecture Notes in Computer Science **2140** (2001) 200-210
24. Rudra, A., Dubey, P., Julta, C., Kumar, V., Rao, J., Rohatgi, P.: Efficient Rijndael implementation with composite field arithmetic. Proc. Cryptographic Hardware and Embedded Systems – CHES'01. Lecture Notes in Computer Science **2162** (2001) 175-188
25. Satoh, A., Morioka, S., Takano, K., Munetoh, S.: A compact Rijndael hardware architecture with S-Box optimization. Proc. Advances in Cryptology – ASIACRYPT 2001. Lecture Notes in Computer Science **2248** (2001) 239-254
26. Tiri, K., Akmal, M., Verbauwhede, I.: A dynamic and differential CMOS logic with signal independent power consumption to withstand differential power analysis on smart cards. Proc. IEEE 28th Europen Solid-State Circuit Conf. – ESSCIRC'02 (2002)

27. E. Trichina, E., De Seta, D., Germani, L.: Simplified Adaptive Multiplicative Masking for AES and its secure implementation. Proc. Cryptographic Hardware and Embedded Systems: CHES 2002. 2523 of Lecture Notes in Computer Science **2523** (2002) 277-285

28. Trichina, E.: Combinational logic design for AES SubByte transformation on masked data. IACR Cryptology ePrint Archive (2003)

29. Wolkerstorfer, J.,: An ASIC implementation of the AES MixColumn operation, In Proceedings Austrochip 2001 (2001)

30. Wolkerstorfer, J., Oswald, E., Lamberger, M.: An ASIC implementation of the AES S-Boxes. Proc. Topic in Cryptography – CT-RSA 2002. 2271 of Lecture Notes in Computer Science **2271** (2002) 67-78

# Complementation-Like and Cyclic Properties of AES Round Functions

Tri Van Le[1], Rüdiger Sparr[2], Ralph Wernsdorf[2], and Yvo Desmedt[1]

[1] Dept. of Computer Science, Florida State University,
260 Love Building, Tallahassee, FL 32306-4530, USA
tl16935@garnet.acns.fsu.edu
desmedt@cs.fsu.edu
[2] Rohde & Schwarz SIT GmbH,
Agastraße 3, D-12489 Berlin, Germany
{ruediger.sparr, ralph.wernsdorf}@sit.rohde-schwarz.com

**Abstract.** While it is known previously that the cycle lengths of individual components of the AES round function are very small, we demonstrate here that the cycle length of the S-box combined with the ShiftRow and MixColumn transformation is at least $10^{205}$. This result is obtained by providing new invariances of the complete AES round function without the key addition. Furthermore, we consider self-duality properties of the AES round function and derive a property analogous to the complementation property of the DES round function. These results confirm the assessments given in other publications that the AES components have several unexpected structural properties.

**Keywords:** Rijndael, AES, invariance, cyclic properties, self-duality.

## 1 Introduction

The cipher Rijndael was selected in 2000 as the Advanced Encryption Standard (AES) and was designed to resist known attacks to block ciphers up to that time. In particular, Rijndael is considered to be immune to differential cryptanalysis [2] and linear cryptanalysis [6], cf. [10]. On the other hand, in order to achieve a good performance on different platforms, the components of Rijndael operate completely on the Galois field $GF(2^8)$. Some recent algebraic analyses of the AES try to exploit the algebraic structure of the finite field $GF(2^8)$ which lead to algebraic relations and simple algebraic representations (cf. [4], [8], and [15]). The analysis of such mathematical properties can lead to new cryptanalytic insights and approaches. In many cases this seems to be the only way to find cryptographic weaknesses. Regularities of algorithm components, such as short cycles, "inner structures" or symmetries can yield starting points for attacks.

In this paper we present some new results on the algebraic properties of the AES round function. The paper is organized as follows: In Section 2 we first give a short description of the AES round function and fix some definitions and

H. Dobbertin, V. Rijmen, A. Sowa (Eds.): AES 2004, LNCS 3373, pp. 128–141, 2005.
© Springer-Verlag Berlin Heidelberg 2005

notations with respect to invariances and permutation groups. In Section 3 we summarize known results on the cyclic order of basic components of the AES round function and provide some new results. In Section 4 we present a number of new invariances of the AES round function with the all zero subkey. In addition, we have found more invariances of some powers of this round function. From the result proved in [13] that the set of AES round functions generates the alternating group on $\{0, 1\}^{128}$, it follows that the existence of nontrivial invariances which hold for all $2^{128}$ AES round subkeys can be excluded. In Section 5 we compute different cycle lengths of the AES round function under the all zero subkey exploiting the invariances found in Section 4. The cycle computations performed in Section 5 show that the cyclic order of this round function has by far not the small size as the cyclic orders of its components. In Section 6 we consider self-duality properties of the AES round function and we derive a property analogous to the complementation property of the DES round function.

## 2    Preliminaries

### 2.1    Description of the AES Round Function

We now give a short description of the AES round function. The AES is defined for 128-bit blocks and key sizes 128, 192, and 256 bits (cf. [9]). The bytes $b_i$ of the state space of the AES are written in matrix form:

| $b_1$ | $b_5$ | $b_9$ | $b_{13}$ |
|---|---|---|---|
| $b_2$ | $b_6$ | $b_{10}$ | $b_{14}$ |
| $b_3$ | $b_7$ | $b_{11}$ | $b_{15}$ |
| $b_4$ | $b_8$ | $b_{12}$ | $b_{16}$ |

In the following we write the state of the AES as a byte vector of the form $(b_1, b_2, b_3, b_4, ..., b_{16})$ with meaning as a matrix of the form as indicated above. A round of Rijndael proceeds in the following 4 consecutive operations on $\{0, 1\}^{128}$ (cf. [9]):

To every byte of the state the S-Box is applied (*ByteSub*), the bytes of the rows of the state matrix are shifted (*ShiftRow*), every byte-column is mixed by a linear transformation (*MixColumn*), and finally, the state is XORed with the 128-bit subkey (*AddRoundKey*).

The S-box of Rijndael is composed as the inversion in the Galois field $GF(2^8)$ modulo the irreducible polynomial $x^8 + x^4 + x^3 + x + 1$, followed by an affine transformation (see [9] for details). The application of the inversion in $GF(2^8)$, resp. the affine transformation described above to the 16 bytes of the state is denoted by $I$ and $A$, respectively.

In the ShiftRow operation, the bytes of row $i$ of the state are rotated $i$ places to the left, where $i = 0, 1, 2, 3$.

In the MixColumn operation, each column of the state is considered as a polynomial over $GF(2^8)$ and multiplied modulo $x^4 + 1$ with the polynomial $03 \cdot x^3 + 01 \cdot x^2 + 01 \cdot x + 02$.

The ShiftRow, MixColum, AddRoundKey operation is denoted by $S$, $M$, and $X$, respectively.

## 2.2    Definitions and Notation

Let $Y$ be a nonempty finite set and let $G$ be a permutation group on $Y$, i.e., a subgroup of the group of bijective mappings of $Y$ to itself. $G$ is called *transitive* if, for any pair of elements $(y, y') \in Y^2$, there is a permutation $f \in G$ with $f(y) = y'$, otherwise $G$ is said to be *intransitive*. The *cyclic order* of an element $f \in G$ is the cardinality of the cyclic subgroup generated by $f$ or, in other words, the least natural number $n > 0$ such that $f^n(y) = y$ for all $y \in Y$. Further information on the theory of permutation groups can be found in [11] or [14].

Let $f : \{0,1\}^{128} \to \{0,1\}^{128}$ denote an operation on the state space of Rijndael. A property $P \subseteq \{0,1\}^{128}$, $P \neq \emptyset$, is called an *invariance* of $f$, if $P$ is preserved by $f$, i.e., for every $x \in P$ it follows that $f(x) \in P$.

In this paper we use the notation $(f \circ g)(x) = g(f(x))$ for compositions of functions $f, g$. The cardinality of a set $Y$ is denoted by $|Y|$. Furthermore, we write $\mathbb{F}_2$ for the Galois field with two elements.

## 3    Cyclic Order of Components of $I \circ A \circ S \circ M$

In this section we summarize known results on the cyclic order of (compositions of) basic components of the Rijndael round function and provide some new results.

Clearly, the cyclic order of $I$, $S$, $X$, $X \neq$ id, is 2, 4, 2, respectively. Song and Seberry showed that the cyclic order of the MixColumn operation $M$ is 4 (cf. [12]). Because the cycle lengths for $I \circ A$ applied to bytes are 87, 81, 59, 27, 2, the cyclic order of $I \circ A$ applied to the AES state space is the least common multiple of $87, 81, 59, 27, 2$ which is 277182 (cf. [12]). Then it follows that the cyclic order of $I \circ A \circ S$ is 554364 (cf. [12]) which is the least common multiple of 277182 and 4. Murphy and Robshaw [7] showed that the order of $A' \circ S \circ M$ is 16, where $A'(a)(u) = (u^7 + u^6 + u^5 + u^4 + 1)a(u) \bmod u^8 + 1$.

**Proposition 1.** *The cyclic order of the affine transformation $A$ is 4.*

*Proof.* According to [3], the affine transformation $A$ can be written as a mapping of the ring $\mathbb{F}_2[u]/(u^8 + 1)$

$$A : a(u) \mapsto f(u)a(u) + g(u) \bmod u^8 + 1,$$

where $f(u) = u^7 + u^6 + u^5 + u^4 + 1$, and $g(u) = u^7 + u^6 + u^2 + u$. Then we have $f(u)^2 = u^6 + u^4 + u^2 \bmod u^8 + 1$ and $f(u)^4 = 1 \bmod u^8 + 1$. Since $f(u)^3 + f(u)^2 + f(u) = 1 \bmod u^8 + 1$, it follows that $A^4(a)(u) = f(u)^4 a(u) + (f(u)^3 + f(u)^2 + f(u) + 1)g(u) = a(u) \bmod u^8 + 1$.     $\square$

The next two propositions summarize the results about the cyclic orders for step-2 and step-3 compositions of basic components of the AES round function. The cyclic orders of the mappings $I \circ A$ and $I \circ A \circ S$ were stated by Song and Seberry in [12].

**Proposition 2.** *The cyclic order of $I \circ A$, $A \circ S$, $S \circ M$, $M \circ X$ for $X \neq \mathrm{id}$, is 277182, 4, 8, 8, respectively.*

*Proof.* From the preceding proposition it follows that the cyclic order of $A \circ S$ must be 4. Because the $S \circ M$-mapping is $\mathbb{F}_2$-linear, it can be represented by a $128 \times 128$-matrix over $\mathbb{F}_2$, whose order is 8. It remains to prove the result for the cyclic order of $M \circ X$. We consider $M$ as an element in the ring $\mathrm{Mat}_{128}(\mathbb{F}_2)$ of $128 \times 128$-matrices over $\mathbb{F}_2$. Let $E$ denote the unit element of $\mathrm{Mat}_{128}(\mathbb{F}_2)$. Then we have $(M \circ X)^8(y) = M^8 y + (M+E)^7 x$, where $y$ denotes an 128-bit vector over $\mathbb{F}_2$ and $x$ denotes the 128-bit key added by $X$. Since $(M + E)^4 = M^4 + E^4 = 0$, it follows that $(M \circ X)^8(y) = y$ for all 128-bit vectors $y$ over $\mathbb{F}_2$. Because there exist cycles for $M \circ X$ of length 8, the cyclic order of $M \circ X$ is 8. $\qquad\square$

Similar arguments yield the following results for the cyclic order of the mappings $A \circ S \circ M$, $S \circ M \circ X$, and $A \circ S \circ M \circ X$, with $X \neq \mathrm{id}$, stated in Proposition 3 and 4.

**Proposition 3.** *The cyclic order of $I \circ A \circ S$, $A \circ S \circ M$, $S \circ M \circ X$ for $X \neq \mathrm{id}$, is 554364, $\leq 32$, 16, respectively.* $\qquad\square$

**Proposition 4.** *The cyclic order of $A \circ S \circ M \circ X$ for $X \neq \mathrm{id}$ is $\leq 32$.* $\qquad\square$

## 4   Invariances of $I \circ A \circ S \circ M$

In this section we list various invariances of the mapping $I \circ A \circ S \circ M$ and its powers. The following three propositions can easily be verified.

**Proposition 5.** *The following sets $\mathrm{Inv}_1, \ldots, \mathrm{Inv}_6$ are invariances of the mapping $I \circ A \circ S \circ M$.*

$\mathrm{Inv}_1 = \{(x, x, x, \ldots, x) | x \in GF(2^8)\}$,

$\mathrm{Inv}_2 = \{(x, y, x, y, x, y, \ldots, x, y) | x, y \in GF(2^8)\}$,

$\mathrm{Inv}_3 = \{(w, x, y, z, w, x, y, z, w, x, y, z, w, x, y, z) | w, x, y, z \in GF(2^8)\}$,

$\mathrm{Inv}_4 = \{(w, x, w, x, y, z, y, z, w, x, w, x, y, z, y, z) | w, x, y, z \in GF(2^8)\}$,

$\mathrm{Inv}_5 = \{(w, x, y, z, y, z, w, x, w, x, y, z, y, z, w, x) | w, x, y, z \in GF(2^8)\}$,

$\mathrm{Inv}_6 = \{(s, t, u, v, w, x, y, z, s, t, u, v, w, x, y, z) | s, t, u, v, w, x, y, z \in GF(2^8)\}$. $\quad\square$

**Proposition 6.** *The following sets $\mathrm{Inv}_7, \ldots, \mathrm{Inv}_{10}$ are invariances of the mapping $(I \circ A \circ S \circ M)^2$.*

$\mathrm{Inv}_7 = \{(x, x, x, x, y, y, y, y, x, x, x, x, y, y, y, y) | x, y \in GF(2^8)\}$,

$\mathrm{Inv}_8 = \{(x, y, x, y, y, x, y, x, x, y, x, y, y, x, y, x) | x, y \in GF(2^8)\}$,

$\mathrm{Inv}_9 = \{(s, t, s, t, u, v, u, v, w, x, w, x, y, z, y, z) | s, t, u, v, w, x, y, z \in GF(2^8)\}$,

$\mathrm{Inv}_{10} = \{(s, t, u, v, w, x, y, z, u, v, s, t, y, z, w, x) | s, t, u, v, w, x, y, z \in GF(2^8)\}$. $\quad\square$

**Proposition 7.** *The following sets* $\mathrm{Inv}_{11}, ..., \mathrm{Inv}_{14}$ *are invariances of the mapping* $(I \circ A \circ S \circ M)^4$.

$\mathrm{Inv}_{11} = \{(w, w, w, w, x, x, x, x, y, y, y, y, z, z, z, z) | w, x, y, z \in GF(2^8)\},$

$\mathrm{Inv}_{12} = \{(w, x, y, z, x, y, z, w, y, z, w, x, z, w, x, y) | w, x, y, z \in GF(2^8)\},$

$\mathrm{Inv}_{13} = \{(w, x, w, x, y, z, y, z, x, w, x, w, z, y, z, y) | w, x, y, z \in GF(2^8)\},$

$\mathrm{Inv}_{14} = \{(w, x, y, z, z, w, x, y, y, z, w, x, x, y, z, w) | w, x, y, z \in GF(2^8)\}. \qquad \square$

Note that the invariances of $(I \circ A \circ S \circ M)^2$ resp. $(I \circ A \circ S \circ M)^4$ listed in Propositions 6 and 7 are not invariances of $I \circ A \circ S \circ M$.

**Proposition 8.** *There exist no invariances* $P$ *of* $I \circ A \circ S \circ M \circ X$ *such that* $P \neq \{0, 1\}^{128}$ *which hold for all* $2^{128}$ *round subkeys.*

*Proof.* The existence of a nontrivial invariance of $I \circ A \circ S \circ M \circ X$ that holds for all $2^{128}$ round subkeys would imply that the permutation group generated by all $2^{128}$ round functions is intransitive. But this contradicts the fact from [13] that this group is the alternating group on $\{0, 1\}^{128}$. $\qquad \square$

*Remark 1.* The same argument shows that for any natural number $n > 0$ there are no nontrivial invariances $P$ of $(I \circ A \circ S \circ M \circ X)^n$ which hold for all $2^{128 \cdot n}$ combinations of $n$ round subkeys. $\qquad \square$

Nevertheless, for special sets of round functions we obtain the following invariances.

*Remark 2.* (a) For all $i \in \{1, 2, ..., 6\}$ we have: If the round subkey is an element of $\mathrm{Inv}_i$, then $\mathrm{Inv}_i$ is an invariance of $I \circ A \circ S \circ M \circ X$.
(b) For $i \in \{7, 8, ..., 14\}$ similar properties can be derived by a suitable choice of the round keys over 2 and 4 rounds respectively. $\qquad \square$

This means that for round subkeys and input blocks with one of the structures given above, the round function of AES has some strong regularities. From the size of the invariance $\mathrm{Inv}_6$, we obtain the following result.

**Corollary 1.** *There exists a set of AES round functions* $\mathcal{R}$ *with* $|\mathcal{R}| = 2^{64}$ *such that the permutation group generated by* $\mathcal{R}$ *is intransitive.* $\qquad \square$

One cannot expect that these invariances can be extended to the complete AES because the AES key scheduling is designed to avoid regularities.

## 5   On the Cyclic Order of $I \circ A \circ S \circ M$

Because the mappings $I$, $A$, $S$, $M$ and their concatenations $I \circ A$, $A \circ S$, $S \circ M$, $I \circ A \circ S$, and $A \circ S \circ M$ all have relatively small cyclic orders, it is important to check whether this causes small cyclic orders of the AES round functions.

**Table 1.** Cycle Lengths for Invariances $Inv_1, ..., Inv_5$

| Invariance: | Cycle lengths: |
|---|---|
| $Inv_1$ | 87, 81, 59, 27, 2 |
| $Inv_2 \setminus Inv_1$ | 39488, 16934, 7582, 548, 36, 24, 21, 15, 8, 2 |
| $Inv_3 \setminus Inv_2$ | 1088297796, 637481159, 129021490, 64376666, 11782972, 13548, 10756, 5640, 3560, 1902, 136, 90, 47, 40, 12, 4 |
| $Inv_4 \setminus Inv_2$ | 1219717400, 599556416, 315637164, 4307366, 2990738, 2153683, 1958224, 1606154, 1495369, 975150, 803077, 564988, 487575, 86038, 82750, 67324, 21758, 13024, 10902, 5451, 5354, 3340, 2677, 2356, 988, 856, 108, 48, 22, 20, 18, 12, 11, 9, 8, 4 |
| $Inv_5 \setminus Inv_2$ | 1052651234, 737292504, 417828286, 193225414, 96612707, 87601912, 11068518, 9460050, 6486298, 5534259, 4730025, 3243149, 394266, 197133, 42454, 16932, 3166, 3078, 2366, 1583, 1539, 1183, 1160, 1062, 912, 496, 38, 18, 9, 6, 2 |

The round function $I \circ A \circ S \circ M$ with the all zero round subkey is intuitively one of the first candidates for which a small cyclic order seems to be possible. The results of Section 4 can be exploited to find out a lower bound for the cyclic order of this round function. Consider the invariance sets of the permutation $I \circ A \circ S \circ M$ given in Section 4. Since a cycle starting from a point of an invariance set can only contain points of this set, the cycle length is limited by the size of the invariance set. Because we have $|Inv_i| \leq 2^{32}$ for $i = 1, ..., 5$, it is possible to compute such cycles on a PC. Table 1 lists all cycle lengths according to the invariances $Inv_1, ..., Inv_5$. Note that the cycles from the invariance $Inv_3$ are related to those provided in the Appendix of [12]. The Appendix provides a complete listing of the cycles for $Inv_1, ..., Inv_5$.

The least common multiple of the computed cycle lengths for the invariances $Inv_1, ..., Inv_5$ of Table 1 is equal to

15480 20902 25688 03988 20263 80165 33732 81646 22636
91465 18521 79549 12467 08119 71956 75745 56484 49918
71194 74949 82013 75604 65431 85505 89291 91969 54985
57959 09774 73220 08631 61663 54665 95490 84924 52493
78665 19158 83139 45332 15200 0,

which is a number of 206 decimals. Since the least common multiple of the cycle lengths for $Inv_1, ..., Inv_5$ is a divisor of the cyclic order of $I \circ A \circ S \circ M$, we obtain the following result.

**Proposition 9.** *The cyclic order of $I \circ A \circ S \circ M$ is greater than $10^{205}$.*    □

Starting from other points in other invariance sets still other cycles can be found. This way the lower bound can be further increased essentially.

It follows that the cyclic order of $I \circ A \circ S \circ M$ has by far not the small size as the cyclic orders of its components, but for points from invariance sets short cycles may occur. Furthermore, the cyclic order of $I \circ A \circ S \circ M$ is much greater than the number $2^{128}$ of AES blocks.

## 6    Self-duality of the AES Round Function

According to [1], two block ciphers $E$ and $E'$ are called *dual* to each other if there are invertible transformations $f$, $g$, and $h$ such that

$$f(E_k(x)) = E'_{g(k)}(h(x))$$

holds for all keys $k$ and plaintexts $x$. Any cipher is trivially dual to itself, but sometimes there are *nontrivial* invertible transformations for a cipher onto itself such that the equation above holds for all $k$ and $x$, as is the case for the DES (cf. [5], p. 248). In [1] the question is considered whether the AES block encryption has self-duality properties, i.e.:

Do nontrivial invertible transformations $f$, $g$, $h$ exist such that for the AES block encryption for all plaintext blocks $x$ and all keys $k$ the equation $f(AES_k(x)) = AES_{g(k)}(h(x))$ holds ?

In this section we show that this question has a positive answer according to the *AES round function*.

For any natural number $n > 0$, let $S_n$ denote the symmetric group of degree $n$, i.e., the permutation group on the set $\{1, ..., n\}$. Now we set $\pi_0 := (1\ 6\ 11\ 16)$, $\pi_1 := (5\ 10\ 15\ 4)$, $\pi_2 := (9\ 14\ 3\ 8)$, and $\pi_3 := (13\ 2\ 7\ 12)$, and define $G$ as the semidirect product of the group generated by $\pi_0, \pi_1, \pi_2, \pi_3$ and $S_4$, where $S_4$ operates arbitrarily on the four vectors $v_0 := (1, 6, 11, 16)$, $v_1 := (5, 10, 15, 4)$, $v_2 := (9, 14, 3, 8)$, and $v_3 := (13, 2, 7, 12)$. Then we obtain the following result.

**Proposition 10.** *There exists a permutation group $G$ of order $|G| = 6144$ such that for any byte-permutation $\pi \in G$, there exists a byte-permutation $\pi'$ such that $\forall x \in \{0, 1\}^{128} : \pi \circ I \circ A \circ S \circ M(x) = I \circ A \circ S \circ M \circ \pi'(x)$.*    □

Although one cannot expect that this property can be extended to the complete AES mapping, the property of Proposition 10 opens some new possibilities for protection against side channel attacks. Furthermore, Proposition 10 is related to the results provided in Appendix A of [15] where the expressions of the 128 bit-components of the AES round function have many similarities and partially the same component expressions.

On the basis of Proposition 10 it is possible to derive self-duality properties of the AES round function. If $X$ is added, we obtain the following result.

**Corollary 2.** *There exists a permutation group $G$ of order $|G| = 6144$ such that for any byte-permutation $\pi \in G$, there exists a byte-permutation $\pi'$ such that $(\pi \circ I \circ A \circ S \circ M \circ X(k))(x) = (I \circ A \circ S \circ M \circ X(\pi'^{-1}(k)) \circ \pi')(x)$ holds for all $k \in \{0,1\}^{128}$ and $x \in \{0,1\}^{128}$.*     □

If we consider in Corollary 2 the special case $\pi = \pi'$, then we find the following byte permutations (written as products of cycles, the bytes are enumerated as described in Section 2.1).

$$P_1 = (1\ 5\ 9\ 13)(2\ 6\ 10\ 14)(3\ 7\ 11\ 15)(4\ 8\ 12\ 16),$$
$$P_2 = (1\ 9)(5\ 13)(2\ 10)(6\ 14)(3\ 11)(7\ 15)(4\ 12)(8\ 16),$$
$$P_3 = (1\ 13\ 9\ 5)(2\ 14\ 10\ 6)(3\ 15\ 11\ 7)(4\ 16\ 12\ 8).$$

This means that the following automorphism equations hold for the AES round function.

**Proposition 11.** *For any byte permutation $P_i$, $i = 1,2,3$ defined above, the equation $P_i((I \circ A \circ S \circ M \circ X(k))(x)) = (I \circ A \circ S \circ M \circ X(P_i(k)))(P_i(x))$ holds for all $k \in \{0,1\}^{128}$ and $x \in \{0,1\}^{128}$.*     □

This way we have found a property analogous to the complementation property of the DES round function (see for example p. 248 in [5]).

## 7    Conclusions

Novel invariances of the mapping $I \circ A \circ S \circ M$ were found. It follows that for big sets of round subkeys there are regularities in the corresponding round functions. On the other hand, the existence of nontrivial invariances which hold for all $2^{128}$ AES round subkeys is excluded.

Several cycle lengths for the complete AES round function with the all zero subkey were computed. It turns out that the cyclic order of this round function is much greater than the cyclic orders of its components. Nevertheless, for several special round keys some short cycles exist.

The self-duality properties described in Section 6 confirm the assessments given in other publications that the AES components have several unexpected structural properties.

We conclude that the AES has many algebraic properties which have not been found before in other block ciphers. In particular, the round function seems to have more invariants than the round function of DES. The results are not necessarily suitable to break AES. But in combination with other approaches they may lead to new insights and analysis methods.

## References

1. E. Barkan and E. Biham. In how many ways can you write Rijndael? In *Advances in Cryptology - ASIACRYPT 2002*, Lecture Notes in Computer Science 2501, pp. 160-175, Springer-Verlag, 2002.

2. E. Biham and A. Shamir. Differential cryptanalysis of DES-like cryptosystems. *J. Cryptology*, Vol. 4, pp. 3-72, 1991.

3. J. Daemen and V. Rijmen. *AES Proposal: Rijndael.* Available via http://csrc.nist.gov/CryptoToolkit/aes, September 3, 1999.

4. N. Ferguson, R. Schroeppel, and D. Whiting. A simple algebraic representation of Rijndael. In *Selected Areas in Cryptography, SAC 2001*, Lecture Notes in Computer Science 2259, pp. 103-111, Springer-Verlag, 2001.

5. A. G. Konheim. *Cryptography: A Primer.* John Wiley and Sons, 1981.

6. M. Matsui. Linear cryptanalysis method for DES cipher. In *Advances in Cryptology - EUROCRYPT'93*, Lecture Notes in Computer Science 765, pp. 386-397, Springer-Verlag, 1994.

7. S. Murphy and M. J. B. Robshaw. New observations on Rijndael. Available via http://csrc.nist.gov/CryptoToolkit/aes, August 7, 2000.

8. S. Murphy and M. J. B. Robshaw. Essential algebraic structure within the AES. In *Advances in Cryptology - CRYPTO 2002*, Lecture Notes in Computer Science 2442, pp. 1-16, Springer-Verlag, 2002.

9. National Institute of Standards and Technology (U.S.): *Advanced Encryption Standard (AES)*, FIPS Publication 197, November 26, 2001. Available at http://csrc.nist.gov/publications/fips/fips197/ fips-197.pdf.

10. S. Park, S. H. Sung, S. Lee, and J. Lim. Improving the upper bound on the maximum differential and the maximum linear hull probability for SPN structures and AES. In *Fast Software Encryption, 10th International Workshop, FSE 2003*, Lecture Notes in Computer Science 2887, pp. 247-260, Springer-Verlag, 2003.

11. D. Robinson. *A Course in the Theory of Groups.* Graduate Texts in Mathematics, Springer-Verlag, New York, 1982.

12. B. Song and J. Seberry. Further observations on the structure of the AES algorithm. In *Fast Software Encryption, 10th International Workshop, FSE 2003*, Lecture Notes in Computer Science 2887, pp. 223-234, Springer-Verlag, 2003.

13. R. Wernsdorf. The round functions of Rijndael generate the alternating group. In *Fast Software Encryption, 9th International Workshop, FSE 2002*, Lecture Notes in Computer Science 2365, pp. 143-148, Springer-Verlag, 2002.

14. H. Wielandt. *Finite Permutation Groups.* Academic Press, New York, 1964.

15. A. M. Youssef and S. E. Tavares. On some algebraic structures in the AES round function, http://eprint.iacr.org/2002/144, September 20, 2002.

# A    Some Cycles of $I \circ A \circ S \circ M$

The following tables provide a complete listing of the cycles for the invariances $\text{Inv}_1, ..., \text{Inv}_5$.

## A.1    Cycles of $I \circ A \circ S \circ M$ for $\text{Inv}_1$

| | Starting point in INV$_1$: | Cycle length: |
|---|---|---|
| 1 | $(04_x, 04_x, 04_x, 04_x, 04_x, 04_x, 04_x, 04_x, ...\ )$ | 87 |
| 2 | $(01_x, 01_x, 01_x, 01_x, 01_x, 01_x, 01_x, 01_x, ...\ )$ | 81 |
| 3 | $(00_x, 00_x, 00_x, 00_x, 00_x, 00_x, 00_x, 00_x, ...\ )$ | 59 |
| 4 | $(0b_x, 0b_x, 0b_x, 0b_x, 0b_x, 0b_x, 0b_x, 0b_x, ...)$ | 27 |
| 5 | $(73_x, 73_x, 73_x, 73_x, 73_x, 73_x, 73_x, 73_x, ...\ )$ | 2 |

## A.2    Cycles of $I \circ A \circ S \circ M$ for $\text{Inv}_2 \setminus \text{Inv}_1$

| | Starting point in INV$_2$ \ Inv$_1$: | Cycle length: |
|---|---|---|
| 1 | $(00_x, 02_x, 00_x, 02_x, 00_x, 02_x, 00_x, 02_x, ...\ )$ | 39488 |
| 2 | $(00_x, 01_x, 00_x, 01_x, 00_x, 01_x, 00_x, 01_x, ...\ )$ | 16934 |
| 3 | $(00_x, 07_x, 00_x, 07_x, 00_x, 07_x, 00_x, 07_x, ...\ )$ | 7582 |
| 4 | $(00_x, b8_x, 00_x, b8_x, 00_x, b8_x, 00_x, b8_x, ...)$ | 548 |
| 5 | $(00_x, c6_x, 00_x, c6_x, 00_x, c6_x, 00_x, c6_x, ...\ )$ | 548 |
| 6 | $(03_x, d6_x, 03_x, d6_x, 03_x, d6_x, 03_x, d6_x, ...)$ | 36 |
| 7 | $(07_x, f1_x, 07_x, f1_x, 07_x, f1_x, 07_x, f1_x, ...\ )$ | 36 |
| 8 | $(03_x, d5_x, 03_x, d5_x, 03_x, d5_x, 03_x, d5_x, ...)$ | 24 |
| 9 | $(05_x, 0f_x, 05_x, 0f_x, 05_x, 0f_x, 05_x, 0f_x, ...\ )$ | 21 |
| 10 | $(0f_x, 05_x, 0f_x, 05_x, 0f_x, 05_x, 0f_x, 05_x, ...\ )$ | 21 |
| 11 | $(06_x, 86_x, 06_x, 86_x, 06_x, 86_x, 06_x, 86_x, ...\ )$ | 15 |
| 12 | $(0e_x, 6e_x, 0e_x, 6e_x, 0e_x, 6e_x, 0e_x, 6e_x, ...\ )$ | 15 |
| 13 | $(2d_x, 4a_x, 2d_x, 4a_x, 2d_x, 4a_x, 2d_x, 4a_x, ...)$ | 8 |
| 14 | $(5d_x, a3_x, 5d_x, a3_x, 5d_x, a3_x, 5d_x, a3_x, ...)$ | 2 |
| 15 | $(86_x, c0_x, 86_x, c0_x, 86_x, c0_x, 86_x, c0_x, ...\ )$ | 2 |

## A.3    Cycles of $I \circ A \circ S \circ M$ for $\text{Inv}_3 \setminus \text{Inv}_2$

| | Starting point in Inv$_3$ \ Inv$_2$: | Cycle length: |
|---|---|---|
| 1 | $(00_x, 00_x, 00_x, 03_x, 00_x, 00_x, 00_x, 03_x, ...)$ | 1088297796 |
| 2 | $(00_x, 00_x, 00_x, 02_x, 00_x, 00_x, 00_x, 02_x, ...)$ | 637481159 |
| 3 | $(00_x, 00_x, 00_x, 04_x, 00_x, 00_x, 00_x, 04_x, ...)$ | 637481159 |
| 4 | $(00_x, 00_x, 00_x, 06_x, 00_x, 00_x, 00_x, 06_x, ...)$ | 637481159 |
| 5 | $(00_x, 00_x, 00_x, 08_x, 00_x, 00_x, 00_x, 08_x, ...)$ | 637481159 |
| 6 | $(00_x, 00_x, 00_x, 01_x, 00_x, 00_x, 00_x, 01_x, ...)$ | 129021490 |
| 7 | $(00_x, 00_x, 00_x, 07_x, 00_x, 00_x, 00_x, 07_x, ...)$ | 129021490 |
| 8 | $(00_x, 00_x, 00_x, 09_x, 00_x, 00_x, 00_x, 09_x, ...)$ | 129021490 |
| 9 | $(00_x, 00_x, 00_x, 10_x, 00_x, 00_x, 00_x, 10_x, ...)$ | 129021490 |

| | Starting point in $Inv_3 \setminus Inv_2$: | Cycle length: |
|---|---|---|
| 10 | $(00_x, 00_x, 00_x, 16_x, 00_x, 00_x, 00_x, 16_x, \dots )$ | 64376666 |
| 11 | $(00_x, 00_x, 01_x, 42_x, 00_x, 00_x, 01_x, 42_x, \dots )$ | 64376666 |
| 12 | $(00_x, 00_x, 00_x, ea_x, 00_x, 00_x, 00_x, ea_x, \dots )$ | 11782972 |
| 13 | $(00_x, 02_x, 3a_x, f9_x, 00_x, 02_x, 3a_x, f9_x, \dots )$ | 13548 |
| 14 | $(00_x, 05_x, fd_x, e6_x, 00_x, 05_x, fd_x, e6_x, \dots )$ | 13548 |
| 15 | $(00_x, 10_x, 04_x, ad_x, 00_x, 10_x, 04_x, ad_x, \dots )$ | 10756 |
| 16 | $(00_x, 02_x, 2d_x, b0_x, 00_x, 02_x, 2d_x, b0_x, \dots )$ | 5640 |
| 17 | $(00_x, 15_x, e1_x, 86_x, 00_x, 15_x, e1_x, 86_x, \dots )$ | 5640 |
| 18 | $(00_x, 09_x, 40_x, 90_x, 00_x, 09_x, 40_x, 90_x, \dots )$ | 3560 |
| 19 | $(00_x, 00_x, c2_x, 2b_x, 00_x, 00_x, c2_x, 2b_x, \dots )$ | 1902 |
| 20 | $(00_x, 21_x, e4_x, f9_x, 00_x, 21_x, e4_x, f9_x, \dots )$ | 1902 |
| 21 | $(01_x, d2_x, 66_x, c5_x, 01_x, d2_x, 66_x, c5_x, \dots )$ | 136 |
| 22 | $(03_x, 04_x, c1_x, ca_x, 03_x, 04_x, c1_x, ca_x, \dots )$ | 90 |
| 23 | $(02_x, 33_x, 8d_x, 7f_x, 02_x, 33_x, 8d_x, 7f_x, \dots )$ | 90 |
| 24 | $(01_x, 12_x, dc_x, 34_x, 01_x, 12_x, dc_x, 34_x, \dots )$ | 47 |
| 25 | $(01_x, 8b_x, 9d_x, ed_x, 01_x, 8b_x, 9d_x, ed_x, \dots )$ | 47 |
| 26 | $(02_x, 4d_x, b4_x, b1_x, 02_x, 4d_x, b4_x, b1_x, \dots )$ | 47 |
| 27 | $(03_x, c9_x, 75_x, a2_x, 03_x, c9_x, 75_x, a2_x, \dots )$ | 47 |
| 28 | $(0a_x, ff_x, 4a_x, df_x, 0a_x, ff_x, 4a_x, df_x, \dots )$ | 40 |
| 29 | $(03_x, 27_x, 26_x, 6c_x, 03_x, 27_x, 26_x, 6c_x, \dots )$ | 12 |
| 30 | $(01_x, 82_x, 8f_x, c8_x, 01_x, 82_x, 8f_x, c8_x, \dots )$ | 4 |
| 31 | $(27_x, aa_x, 2f_x, 56_x, 27_x, aa_x, 2f_x, 56_x, \dots )$ | 4 |
| 32 | $(7d_x, ad_x, f5_x, a3_x, 7d_x, ad_x, f5_x, a3_x, \dots )$ | 4 |

## A.4    Cycles of $I \circ A \circ S \circ M$ for $Inv_4 \setminus Inv_2$

| | Starting point in $Inv_4 \setminus Inv_2$: | Cycle length: |
|---|---|---|
| 1 | $(00_x, 00_x, 00_x, 00_x, 00_x, 01_x, 00_x, 01_x, \dots )$ | 1219717400 |
| 2 | $(00_x, 00_x, 00_x, 00_x, 00_x, 0a_x, 00_x, 0a_x, \dots )$ | 1219717400 |
| 3 | $(00_x, 00_x, 00_x, 00_x, 00_x, 06_x, 00_x, 06_x, \dots )$ | 599556416 |
| 4 | $(00_x, 00_x, 00_x, 00_x, 00_x, 0e_x, 00_x, 0e_x, \dots )$ | 599556416 |
| 5 | $(00_x, 00_x, 00_x, 00_x, 00_x, 03_x, 00_x, 03_x, \dots )$ | 315637164 |
| 6 | $(00_x, 00_x, 00_x, 00_x, 00_x, 07_x, 00_x, 07_x, \dots )$ | 315637164 |
| 7 | $(00_x, 00_x, 00_x, 00_x, 02_x, 62_x, 02_x, 62_x, \dots )$ | 4307366 |
| 8 | $(00_x, 00_x, 00_x, 00_x, 02_x, 3c_x, 02_x, 3c_x, \dots )$ | 2990738 |
| 9 | $(00_x, 00_x, 00_x, 00_x, 00_x, 16_x, 00_x, 16_x, \dots )$ | 2153683 |
| 10 | $(00_x, 00_x, 00_x, 00_x, 0d_x, ae_x, 0d_x, ae_x, \dots )$ | 2153683 |
| 11 | $(00_x, 00_x, 00_x, 00_x, 01_x, 64_x, 01_x, 64_x, \dots )$ | 1958224 |
| 12 | $(00_x, 00_x, 00_x, 00_x, 01_x, 88_x, 01_x, 88_x, \dots )$ | 1958224 |
| 13 | $(00_x, 00_x, 00_x, 00_x, 07_x, 0f_x, 07_x, 0f_x, \dots )$ | 1606154 |
| 14 | $(00_x, 00_x, 00_x, 00_x, 08_x, 42_x, 08_x, 42_x, \dots )$ | 1495369 |
| 15 | $(00_x, 00_x, 00_x, 00_x, 0e_x, 98_x, 0e_x, 98_x, \dots )$ | 1495369 |
| 16 | $(00_x, 00_x, 00_x, 00_x, 0d_x, e8_x, 0d_x, e8_x, \dots )$ | 975150 |
| 17 | $(00_x, 00_x, 00_x, 00_x, 01_x, 14_x, 01_x, 14_x, \dots )$ | 803077 |
| 18 | $(00_x, 00_x, 00_x, 00_x, 01_x, ad_x, 01_x, ad_x, \dots )$ | 803077 |
| 19 | $(00_x, 00_x, 00_x, 00_x, 05_x, b0_x, 05_x, b0_x, \dots )$ | 564988 |

| | Starting point in $\text{Inv}_4 \setminus \text{Inv}_2$: | Cycle length: |
|---|---|---|
| 20 | $(00_x, 00_x, 00_x, 00_x, 0b_x, 2f_x, 0b_x, 2f_x, \ldots\ )$ | 487575 |
| 21 | $(00_x, 00_x, 00_x, 00_x, 21_x, 9a_x, 21_x, 9a_x, \ldots\ )$ | 487575 |
| 22 | $(00_x, 01_x, 00_x, 01_x, 3c_x, ec_x, 3c_x, ec_x, \ldots\ )$ | 86038 |
| 23 | $(00_x, 01_x, 00_x, 01_x, b9_x, 14_x, b9_x, 14_x, \ldots\ )$ | 86038 |
| 24 | $(00_x, 01_x, 00_x, 01_x, cd_x, a4_x, cd_x, a4_x, \ldots\ )$ | 86038 |
| 25 | $(00_x, 02_x, 00_x, 02_x, a9_x, bb_x, a9_x, bb_x, \ldots\ )$ | 86038 |
| 26 | $(00_x, 00_x, 00_x, 00_x, 06_x, 15_x, 06_x, 15_x, \ldots\ )$ | 82750 |
| 27 | $(00_x, 00_x, 00_x, 00_x, 15_x, 06_x, 15_x, 06_x, \ldots\ )$ | 82750 |
| 28 | $(00_x, 00_x, 00_x, 00_x, 34_x, 48_x, 34_x, 48_x, \ldots\ )$ | 82750 |
| 29 | $(00_x, 01_x, 00_x, 01_x, bc_x, 72_x, bc_x, 72_x, \ldots\ )$ | 82750 |
| 30 | $(00_x, 00_x, 00_x, 00_x, 02_x, 02_x, 02_x, 02_x, \ldots\ )$ | 67324 |
| 31 | $(00_x, 00_x, 00_x, 00_x, 05_x, 05_x, 05_x, 05_x, \ldots\ )$ | 21758 |
| 32 | $(00_x, 00_x, 00_x, 00_x, 17_x, 17_x, 17_x, 17_x, \ldots\ )$ | 21758 |
| 33 | $(00_x, 00_x, 00_x, 00_x, 03_x, 03_x, 03_x, 03_x, \ldots\ )$ | 13024 |
| 34 | $(00_x, 0a_x, 00_x, 0a_x, 63_x, 1e_x, 63_x, 1e_x, \ldots\ )$ | 10902 |
| 35 | $(00_x, 0f_x, 00_x, 0f_x, 79_x, 89_x, 79_x, 89_x, \ldots\ )$ | 5451 |
| 36 | $(00_x, 07_x, 00_x, 07_x, aa_x, 22_x, aa_x, 22_x, \ldots\ )$ | 5451 |
| 37 | $(00_x, 0e_x, 00_x, 0e_x, 4b_x, 2e_x, 4b_x, 2e_x, \ldots\ )$ | 5354 |
| 38 | $(00_x, 00_x, 00_x, 00_x, 01_x, 01_x, 01_x, 01_x, \ldots\ )$ | 3340 |
| 39 | $(00_x, 05_x, 00_x, 05_x, 7f_x, 04_x, 7f_x, 04_x, \ldots\ )$ | 2677 |
| 40 | $(00_x, 29_x, 00_x, 29_x, 8e_x, b1_x, 8e_x, b1_x, \ldots\ )$ | 2677 |
| 41 | $(00_x, 00_x, 00_x, 00_x, 91_x, 91_x, 91_x, 91_x, \ldots\ )$ | 2356 |
| 42 | $(00_x, 0b_x, 00_x, 0b_x, 2e_x, 35_x, 2e_x, 35_x, \ldots\ )$ | 988 |
| 43 | $(00_x, 00_x, 00_x, 00_x, 09_x, 09_x, 09_x, 09_x, \ldots\ )$ | 856 |
| 44 | $(02_x, bb_x, 02_x, bb_x, bb_x, 02_x, bb_x, 02_x, \ldots\ )$ | 108 |
| 45 | $(00_x, 2b_x, 00_x, 2b_x, 38_x, bf_x, 38_x, bf_x, \ldots\ )$ | 48 |
| 46 | $(00_x, 6e_x, 00_x, 6e_x, c2_x, 78_x, c2_x, 78_x, \ldots\ )$ | 48 |
| 47 | $(00_x, ba_x, 00_x, ba_x, 24_x, b5_x, 24_x, b5_x, \ldots\ )$ | 48 |
| 48 | $(13_x, 18_x, 13_x, 18_x, 78_x, 69_x, 78_x, 69_x, \ldots\ )$ | 48 |
| 49 | $(12_x, 13_x, 12_x, 13_x, cd_x, d6_x, cd_x, d6_x, \ldots\ )$ | 22 |
| 50 | $(05_x, f1_x, 05_x, f1_x, 83_x, e9_x, 83_x, e9_x, \ldots\ )$ | 20 |
| 51 | $(0d_x, ea_x, 0d_x, ea_x, 62_x, d1_x, 62_x, d1_x, \ldots\ )$ | 20 |
| 52 | $(07_x, 6c_x, 07_x, 6c_x, 7b_x, 4a_x, 7b_x, 4a_x, \ldots\ )$ | 18 |
| 53 | $(03_x, b4_x, 03_x, b4_x, b4_x, 03_x, b4_x, 03_x, \ldots\ )$ | 12 |
| 54 | $(06_x, 06_x, 06_x, 06_x, 35_x, 35_x, 35_x, 35_x, \ldots\ )$ | 12 |
| 55 | $(05_x, cb_x, 05_x, cb_x, 51_x, ce_x, 51_x, ce_x, \ldots\ )$ | 11 |
| 56 | $(21_x, 7e_x, 21_x, 7e_x, 21_x, eb_x, 21_x, eb_x, \ldots\ )$ | 11 |
| 57 | $(1a_x, 27_x, 1a_x, 27_x, 49_x, fa_x, 49_x, fa_x, \ldots\ )$ | 9 |
| 58 | $(28_x, 8d_x, 28_x, 8d_x, 90_x, ea_x, 90_x, ea_x, \ldots\ )$ | 9 |
| 59 | $(03_x, 2d_x, 03_x, 2d_x, 09_x, 76_x, 09_x, 76_x, \ldots\ )$ | 8 |
| 60 | $(2c_x, 26_x, 2c_x, 26_x, 97_x, 84_x, 97_x, 84_x, \ldots\ )$ | 8 |
| 61 | $(2c_x, b1_x, 2c_x, b1_x, b1_x, 2c_x, b1_x, 2c_x, \ldots\ )$ | 8 |
| 62 | $(12_x, 12_x, 12_x, 12_x, f9_x, f9_x, f9_x, f9_x, \ldots\ )$ | 4 |
| 63 | $(39_x, 53_x, 39_x, 53_x, ea_x, 4b_x, ea_x, 4b_x, \ldots\ )$ | 4 |
| 64 | $(39_x, 67_x, 39_x, 67_x, 9f_x, e4_x, 9f_x, e4_x, \ldots\ )$ | 4 |

## A.5    Cycles of $I \circ A \circ S \circ M$ for $\mathrm{Inv}_5 \setminus \mathrm{Inv}_2$

| | Starting point in $\mathrm{Inv}_5 \setminus \mathrm{Inv}_2$: | Cycle length: |
|---|---|---|
| 1 | $(00_x, 00_x, 00_x, 03_x, 00_x, 03_x, 00_x, 00_x, \ldots)$ | 1052651234 |
| 2 | $(00_x, 00_x, 00_x, 05_x, 00_x, 05_x, 00_x, 00_x, \ldots)$ | 1052651234 |
| 3 | $(00_x, 00_x, 00_x, 04_x, 00_x, 04_x, 00_x, 00_x, \ldots)$ | 737292504 |
| 4 | $(00_x, 00_x, 00_x, 01_x, 00_x, 01_x, 00_x, 00_x, \ldots)$ | 417828286 |
| 5 | $(00_x, 00_x, 00_x, 02_x, 00_x, 02_x, 00_x, 00_x, \ldots)$ | 417828286 |
| 6 | $(00_x, 00_x, 00_x, 2c_x, 00_x, 2c_x, 00_x, 00_x, \ldots)$ | 193225414 |
| 7 | $(00_x, 00_x, 00_x, 18_x, 00_x, 18_x, 00_x, 00_x, \ldots)$ | 96612707 |
| 8 | $(00_x, 00_x, 00_x, 47_x, 00_x, 47_x, 00_x, 00_x, \ldots)$ | 96612707 |
| 9 | $(00_x, 00_x, 00_x, 0e_x, 00_x, 0e_x, 00_x, 00_x, \ldots)$ | 87601912 |
| 10 | $(00_x, 00_x, 00_x, 66_x, 00_x, 66_x, 00_x, 00_x, \ldots)$ | 87601912 |
| 11 | $(00_x, 00_x, 00_x, 49_x, 00_x, 49_x, 00_x, 00_x, \ldots)$ | 11068518 |
| 12 | $(00_x, 00_x, 03_x, e9_x, 03_x, e9_x, 00_x, 00_x, \ldots)$ | 9460050 |
| 13 | $(00_x, 00_x, 04_x, 45_x, 04_x, 45_x, 00_x, 00_x, \ldots)$ | 6486298 |
| 14 | $(00_x, 00_x, 00_x, 43_x, 00_x, 43_x, 00_x, 00_x, \ldots)$ | 5534259 |
| 15 | $(00_x, 00_x, 0b_x, 61_x, 0b_x, 61_x, 00_x, 00_x, \ldots)$ | 5534259 |
| 16 | $(00_x, 00_x, 00_x, 27_x, 00_x, 27_x, 00_x, 00_x, \ldots)$ | 4730025 |
| 17 | $(00_x, 00_x, 01_x, 2f_x, 01_x, 2f_x, 00_x, 00_x, \ldots)$ | 4730025 |
| 18 | $(00_x, 00_x, 05_x, 23_x, 05_x, 23_x, 00_x, 00_x, \ldots)$ | 3243149 |
| 19 | $(00_x, 00_x, 06_x, 49_x, 06_x, 49_x, 00_x, 00_x, \ldots)$ | 3243149 |
| 20 | $(00_x, 00_x, 40_x, 54_x, 40_x, 54_x, 00_x, 00_x, \ldots)$ | 394266 |
| 21 | $(00_x, 00_x, 1a_x, 8f_x, 1a_x, 8f_x, 00_x, 00_x, \ldots)$ | 197133 |
| 22 | $(00_x, 00_x, 22_x, 91_x, 22_x, 91_x, 00_x, 00_x, \ldots)$ | 197133 |
| 23 | $(00_x, 02_x, 9e_x, 6d_x, 9e_x, 6d_x, 00_x, 02_x, \ldots)$ | 42454 |
| 24 | $(00_x, 06_x, 7a_x, f3_x, 7a_x, f3_x, 00_x, 06_x, \ldots)$ | 42454 |
| 25 | $(00_x, 00_x, b9_x, 2d_x, b9_x, 2d_x, 00_x, 00_x, \ldots)$ | 16932 |
| 26 | $(00_x, 02_x, 00_x, 45_x, 00_x, 45_x, 00_x, 02_x, \ldots)$ | 16932 |
| 27 | $(00_x, 02_x, ca_x, 65_x, ca_x, 65_x, 00_x, 02_x, \ldots)$ | 16932 |
| 28 | $(00_x, 09_x, 9e_x, e9_x, 9e_x, e9_x, 00_x, 09_x, \ldots)$ | 16932 |
| 29 | $(00_x, 16_x, ef_x, 51_x, ef_x, 51_x, 00_x, 16_x, \ldots)$ | 3166 |
| 30 | $(00_x, 02_x, 54_x, 1a_x, 54_x, 1a_x, 00_x, 02_x, \ldots)$ | 3078 |
| 31 | $(00_x, 25_x, c7_x, 5e_x, c7_x, 5e_x, 00_x, 25_x, \ldots)$ | 2366 |
| 32 | $(00_x, 02_x, 9c_x, e1_x, 9c_x, e1_x, 00_x, 02_x, \ldots)$ | 1583 |
| 33 | $(00_x, 13_x, 25_x, 5a_x, 25_x, 5a_x, 00_x, 13_x, \ldots)$ | 1583 |
| 34 | $(00_x, 12_x, 23_x, f1_x, 23_x, f1_x, 00_x, 12_x, \ldots)$ | 1539 |
| 35 | $(00_x, 23_x, 71_x, ae_x, 71_x, ae_x, 00_x, 23_x, \ldots)$ | 1539 |
| 36 | $(00_x, 16_x, a2_x, ea_x, a2_x, ea_x, 00_x, 16_x, \ldots)$ | 1183 |
| 37 | $(00_x, 1d_x, 56_x, 47_x, 56_x, 47_x, 00_x, 1d_x, \ldots)$ | 1183 |
| 38 | $(00_x, 84_x, 4d_x, b4_x, 4d_x, b4_x, 00_x, 84_x, \ldots)$ | 1160 |
| 39 | $(00_x, 03_x, 49_x, 72_x, 49_x, 72_x, 00_x, 03_x, \ldots)$ | 1062 |
| 40 | $(00_x, 09_x, 9d_x, 7a_x, 9d_x, 7a_x, 00_x, 09_x, \ldots)$ | 1062 |
| 41 | $(00_x, 24_x, f5_x, 24_x, f5_x, 24_x, 00_x, 24_x, \ldots)$ | 1062 |
| 42 | $(00_x, 5a_x, 24_x, 44_x, 24_x, 44_x, 00_x, 5a_x, \ldots)$ | 1062 |
| 43 | $(00_x, 0f_x, 39_x, 9b_x, 39_x, 9b_x, 00_x, 0f_x, \ldots)$ | 912 |

| | Starting point in $\text{Inv}_5 \setminus \text{Inv}_2$: | Cycle length: |
|---|---|---|
| 44 | $(00_x, 49_x, cf_x, cc_x, cf_x, cc_x, 00_x, 49_x, \ldots\ )$ | 496 |
| 45 | $(00_x, f8_x, 3f_x, e6_x, 3f_x, e6_x, 00_x, f8_x, \ldots\ )$ | 496 |
| 46 | $(02_x, 9b_x, ad_x, 9e_x, ad_x, 9e_x, 02_x, 9b_x, \ldots\ )$ | 38 |
| 47 | $(03_x, 5b_x, 16_x, d9_x, 16_x, d9_x, 03_x, 5b_x, \ldots\ )$ | 38 |
| 48 | $(03_x, 58_x, 9c_x, 44_x, 9c_x, 44_x, 03_x, 58_x, \ldots\ )$ | 18 |
| 49 | $(25_x, de_x, 99_x, 86_x, 99_x, 86_x, 25_x, de_x, \ldots\ )$ | 9 |
| 50 | $(3b_x, dc_x, 96_x, e9_x, 96_x, e9_x, 3b_x, dc_x, \ldots\ )$ | 9 |
| 51 | $(06_x, 83_x, 45_x, d3_x, 45_x, d3_x, 06_x, 83_x, \ldots\ )$ | 6 |
| 52 | $(3e_x, 99_x, 88_x, ce_x, 88_x, ce_x, 3e_x, 99_x, \ldots\ )$ | 6 |
| 53 | $(08_x, 24_x, f2_x, 16_x, f2_x, 16_x, 08_x, 24_x, \ldots\ )$ | 2 |
| 54 | $(24_x, f2_x, 16_x, 08_x, 16_x, 08_x, 24_x, f2_x, \ldots\ )$ | 2 |
| 55 | $(35_x, ec_x, c0_x, a7_x, c0_x, a7_x, 35_x, ec_x, \ldots\ )$ | 2 |
| 56 | $(c0_x, a7_x, 35_x, ec_x, 35_x, ec_x, c0_x, a7_x, \ldots\ )$ | 2 |

# More Dual Rijndaels

Håvard Raddum

Dep. of Informatics, The University of Bergen, P.O.box 7800, 5020 Bergen, Norway

**Abstract.** It is well known that replacing the irreducible polynomial used in the AES one can produce 240 dual ciphers. In this paper we present 9120 other representations of $GF(2^8)$, producing more ciphers dual to the AES. We also show that if the matrix used in the S-box of Rijndael is linear over a larger field than $GF(2)$, this would have implications for the XSL attack.

## 1   Introduction

The cipher Rijndael [1] has been selected by NIST as the AES. Most of the operations in Rijndael are based on the field $GF(2^8)$, and several researchers have made comments on the algebraic structures found in the cipher [3, 4, 5]. At ASIACRYPT 2002 Barkan and Biham [5] showed that the ciphers produced when changing the polynomial used in AES are duals of Rijndael. In this paper we construct many more duals of the AES.

Also at ASIACRYPT 2002 Courtois and Pieprzyk [6] described a possible attack on the AES, using a large system of equations. We will show that one of the dual ciphers could produce a much smaller system, that should be easier to solve. However, we have checked that the matrix used in the affine transformation in the S-box is not among those which would simplify the system of equations.

At EUROCRYPT 2003 Biryukov *et al.* [7] presented a tool for finding affine equivalent S-boxes. This can be used to find 2040 pairs of affine mappings that can be inserted in the AES, without changing the permutation induced by the cipher. By replacing the field polynomial in the AES with one of the 30 other irreducible polynomials, one is likely to be able to produce as many as 61,200 different versions of the duals of the AES found in [5]. This class can probably be extended using the duals presented here.

In Section 2 we give a brief description of Rijndael, and the definition of a dual cipher. In Section 3 we show how to construct 1170 different representations of $GF(2^8)$, each one resulting in 8 ciphers dual to the AES. In Section 4 we check whether the system of equations in the XSL-attack can be simplified. Conclusions are made in Section 5.

## 2   Description of Rijndael

We here give a brief description of Rijndael, omitting the key schedule. A more detailed description can be found in [1].

H. Dobbertin, V. Rijmen, A. Sowa (Eds.): AES 2004, LNCS 3373, pp. 142–147, 2005.

Rijndael is a 128-bit block cipher with key sizes of 128, 192 or 256 bits. The cipher consists of a round function that is repeated 10, 12 or 14 times according to the length of the key. The cipher block and the round keys are viewed as $4 \times 4$-matrices of bytes. In some operations these bytes are viewed as elements of $GF(2^8)$, as well as 8-bit strings. The irreducible polynomial over $GF(2)$ used to represent $GF(2^8)$ is $x^8 + x^4 + x^3 + x + 1$.

There are four operations in the round function of Rijndael. These are used in the following order:

- SubBytes
- ShiftRows
- MixColumns
- AddRoundKey

SubBytes replaces each byte of the cipher block. Each byte is first replaced by its inverse, when viewed as an element of $GF(2^8)$ $(0^{-1} = 0)$, and then passed through an affine transformation $Ax + b$ as an 8-bit vector. The constants $A$ and $b$ are

$$
A = \begin{pmatrix}
1 & 0 & 0 & 0 & 1 & 1 & 1 & 1 \\
1 & 1 & 0 & 0 & 0 & 1 & 1 & 1 \\
1 & 1 & 1 & 0 & 0 & 0 & 1 & 1 \\
1 & 1 & 1 & 1 & 0 & 0 & 0 & 1 \\
1 & 1 & 1 & 1 & 1 & 0 & 0 & 0 \\
0 & 1 & 1 & 1 & 1 & 1 & 0 & 0 \\
0 & 0 & 1 & 1 & 1 & 1 & 1 & 0 \\
0 & 0 & 0 & 1 & 1 & 1 & 1 & 1
\end{pmatrix}, \qquad
b = \begin{pmatrix}
1 \\ 1 \\ 0 \\ 0 \\ 0 \\ 1 \\ 1 \\ 0
\end{pmatrix}
$$

ShiftRows takes row $i$ of the cipher block, containing four bytes, and shifts it $i$ positions to the left. The top row is row 0 and the bottom is row 3.

MixColumns views the cipher state as a $4 \times 4$-matrix over $GF(2^8)$, and pre-multiplies it with a constant $4 \times 4$-matrix with elements from $GF(2^8)$.

AddRoundKey simply xors the cipher block with the key for the current round.

An AddRoundKey is applied to the plaintext before the first round, and in the last round MixColumns is removed.

## 2.1    Dual Ciphers

We give here the definition of a dual cipher from [5].

**Definition 2.1.** *Two ciphers $E$ and $E'$ are called dual ciphers if there exists invertible transformations $f, g$ and $h$ such that*

$$\forall P, K \qquad f(E_K(P)) = E'_{g(K)}(h(P)).$$

In the case for Rijndael in this paper we will have $f = g = h$. The transformation $f$ will be an isomorphism of $GF(2^8)$ applied on all 16 bytes in the cipher block in parallel.

# 3    Different Representations of $GF(2^8)$

The designers of Rijndael chose the irreducible polynomial $r(x) = x^8 + x^4 + x^3 + x + 1$ to construct $GF(2^8)$. In the following let $\alpha$ be a root of $r(x)$. Elements of $GF(2)[\alpha]$ (all sums and products of elements from $GF(2) \cup \{\alpha\}$) may be written as polynomials in $\alpha$ over $GF(2)$, with degree at most 7. The elements of $GF(2^8)$ are sometimes regarded as 8-bit vectors, with the natural mapping

$$c_7\alpha^7 + \ldots + c_1\alpha + c_0 \longleftrightarrow (c_7, \ldots, c_1, c_0).$$

When an element of $GF(2^8)$ is written as a column vector $c_0$ is at the top and $c_7$ is at the bottom.

## 3.1    Dual Ciphers by Replacing $r(x)$

There are 30 irreducible polynomials of degree 8 over $GF(2)$. As pointed out in [5], we may define $\beta$ to be a root of any one of these polynomials, and construct $GF(2^8) = GF(2)[\beta]$. The isomorphism $\phi$ between $GF(2)[\alpha]$ and $GF(2)[\beta]$ is established when we find a root of $r(x)$ in $GF(2)[\beta]$, and let this root be the image of $\alpha$.

This isomorphism is a linear mapping. Let $M_\phi$ be the $8 \times 8$-matrix over $GF(2)$ whose column $i$ is $\phi(\alpha^i)$, where column 0 is the leftmost column and column 7 is the rightmost column. Then $\phi(a)$ can be computed as $\phi(a) = M_\phi \cdot a$, where $a \in GF(2)[\alpha]$ is written as a column vector.

Denote encryption of plaintext $P$ under key $K$ using Rijndael by $E_K(P)$. Let the cipher we get by replacing all constants in $GF(2^8)$ in Rijndael by their image under $\phi$, and replacing $A$ with $M_\phi A M_\phi^{-1}$ be called $E'$. Then we have the duality [5]:

$$\phi(E_K(P)) = E'_{\phi(K)}(\phi(P)),$$

where we understand $\phi$ to be applied to each of the $GF(2^8)$-elements in the blocks $P, K$ and $E_K(P)$.

Since there are 8 different roots of $r(x)$ in $GF(2^8)$, we get 8 different isomorphisms between $GF(2)[\alpha]$ and each representation of $GF(2^8)$. With 30 irreducible polynomials of degree 8 over $GF(2)$ we therefore get a total of 240 different matrices $M_\phi$.

## 3.2    Other Representations of $GF(2^8)$

There are other ways of constructing $GF(2^8)$ than by using an irreducible polynomial of degree 8 over $GF(2)$. This is shown by the following example.

First we create $GF(2^2) = GF(2)[\beta]$ with $\beta^2 + \beta + 1 = 0$. Then we can make $GF(2^8)$ with $t(x) = x^4 + \beta x^3 + x + (\beta + 1)$, an irreducible polynomial of degree 4 over $GF(2)[\beta]$. Defining $\gamma$ to be a root of $t(x)$, the elements of $GF(2^8)$ can be written as polynomials in $\gamma$ of degree at most 3 with coefficients from $GF(2)[\beta]$. Writing elements of $GF(2^2)$ as polynomials in $\beta$ of degree at most 1 over $GF(2)$, we get a natural mapping between 8-bit strings and elements of $GF(2)[\beta, \gamma]$:

$$(c_7\beta + c_6)\gamma^3 + (c_5\beta + c_4)\gamma^2 + \ldots + (c_1\beta + c_0) \longleftrightarrow (c_7, \ldots, c_0). \tag{1}$$

With this mapping the isomorphism $\phi : GF(2)[\alpha] \longrightarrow GF(2)[\beta, \gamma]$ can now be realized as a matrix-multiplication in the same way as in the single extension case. We find a root of $r(x)$ in $GF(2)[\beta, \gamma]$ and let this element be $\phi(\alpha)$. Then $M_\phi = [1, \phi(\alpha), \phi(\alpha^2), \ldots, \phi(\alpha^7)]$.

## 3.3    All Possible Representations of $GF(2^8)$ Using Irreducible Polynomials

Here we will show that there are 1170 different representations of $GF(2^8)$ using roots from irreducible polynomials. We have the following inclusions of subfields of $GF(2^8)$:

$$GF(2) \subset GF(2^2) \subset GF(2^4) \subset GF(2^8).$$

This induces four different chains of fields starting with $GF(2)$ and ending in $GF(2^8)$, these chains are listed below. The number above an arrow in $GF(2^i) \xrightarrow{n} GF(2^{di})$ means there are $n$ irreducible polynomials of degree $d$ over $GF(2^i)$.

- $GF(2) \xrightarrow{30} GF(2^8)$: 30 representations.
- $GF(2) \xrightarrow{1} GF(2^2) \xrightarrow{60} GF(2^8)$: 60 representations.
- $GF(2) \xrightarrow{3} GF(2^4) \xrightarrow{120} GF(2^8)$: 360 representations.
- $GF(2) \xrightarrow{1} GF(2^2) \xrightarrow{6} GF(2^4) \xrightarrow{120} GF(2^8)$: 720 representations.

Adding the numbers together we get 1170 representations of $GF(2^8)$.

The mapping between 8-bit strings and field elements for the last two chains can be done as follows.

$GF(2) \longrightarrow GF(2^4) \longrightarrow GF(2^8)$: Let $\beta$ be a root of an irreducible polynomial of degree 4 over $GF(2)$, and let $\gamma$ be a root of an irreducible polynomial of degree 2 over $GF(2)[\beta]$. The conversion is then

$$(c_7\beta^3 + \ldots + c_4)\gamma + (c_3\beta^3 + \ldots + c_0) \longleftrightarrow (c_7, \ldots, c_0).$$

$GF(2) \longrightarrow GF(2^2) \longrightarrow GF(2^4) \longrightarrow GF(2^8)$: Let $\beta$ be a root of $x^2 + x + 1$, $\gamma$ a root of an irreducible polynomial of degree 2 over $GF(2)[\beta]$, and $\delta$ a root of an irreducible polynomial of degree 2 over $GF(2)[\beta, \gamma]$. The mapping becomes

$$((c_7\beta + c_6)\gamma + (c_5\beta + c_4))\delta + ((c_3\beta + c_2)\gamma + (c_1\beta + c_0)) \longleftrightarrow (c_7, \ldots, c_0).$$

For each representation there are 8 choices for the element $\phi(\alpha)$. In total we then get $8 \cdot 1170 = 9360$ matrices $M_\phi$ yielding isomorphisms, and so 9360 duals of the AES. We have generated all these matrices, and checked that they are all different (However, it can be shown that there are 60 pairs of matrices $\{M, M'\}$ such that the first 4 columns of $M$ and $M'$ are equal).

It should be noted that the idea of constructing $GF(2^8)$ using two field extensions and applying it to Rijndael is not new. It has been done in [8], for the purpose of making an efficient hardware implementation of inversion in $GF(2^8)$.

# 4    Implications for the XSL-Attack

The XSL attack is described in [6]. The basis of the attack is the fact that the non-linear part of the S-box in Rijndael is inversion in the field $GF(2^8)$. If $X$ is the input to the inversion and $Y$ is the output, we have the relation $XY = 1$ (except for $X = 0$). By writing $X$ as $x_7\alpha^7 + \ldots + x_0$ and $Y$ as $y_7\alpha^7 + \ldots + y_0$, the expression

$$(x_7\alpha^7 + \ldots + x_0)(y_7\alpha^7 + \ldots + y_0) = 0 \cdot \alpha^7 + \ldots + 0 \cdot \alpha + 1$$

will give us 8 quadratic equations in the variables $x_0, \ldots, x_7, y_0, \ldots, y_7$.

## 4.1    Brief Summary of the XSL Attack

At some point in each round, we give variable names to the bits of the cipher block. Since all the operations in Rijndael except the field inversion are linear over $GF(2)$, the input and output of the inversion are linear expressions in these variables. By using the relation of the field inversion described above, we can create an equation system in the key bits and the intermediate ciphertext bits using one known plaintext/ciphertext pair. All of these equations will be quadratic, and for the 128-bit key case the system should define the key uniquely.

The rest of the attack is to try to solve this equation system by creating new equations using multiplication with monomials, and in the end using re-linearization. If the XSL attack works, it is important that it is faster than exhaustive search. One crucial point for the complexity of solving the system is the number of variables it contains, and for the re-linearization, the number of monomials.

## 4.2    Matrix in S-Box $GF(2^2)$-Linear?

Let us assume for a little while that the matrix used in the S-box of Rijndael is linear over $GF(2^2)$. The other linear operations are linear over $GF(2^8)$, and in particular over $GF(2^2)$. This means that Rijndael can be described completely in terms of $GF(2^2)$, it will never be necessary to go down to bit level in any of the operations. Since all the linear operations of Rijndael are $GF(2^2)$-linear, we can make an equation system like the one used in the XSL-attack, but now with variables and coefficients from $GF(2^2)$. Since two and two bits are melted together to form one variable, we will only get half as many variables as in the original system, and only about one fourth of the number of quadratic monomials. Since the number of monomials is significantly smaller in the system over $GF(2^2)$, and since we only have half as many variables, it should be easier to reach the point where re-linearization can be applied.

The number of invertible $8\times8$-matrices over $GF(2)$ is about $2^{62.2}$, and of these only about $2^{31.5}$ are linear over $GF(2^2)$. This means a random invertible $GF(2)$-matrix have a probability of less than $2^{-30}$ of being $GF(2^2)$-linear. A check has indeed verified that the matrix used in the S-box of Rijndael is not $GF(2^2)$-linear, and so the system can not be simplified this way. To our knowledge this is the first time it has been checked whether this matrix is linear over a larger field.

# 5   Conclusions

In this paper we have increased the list of ciphers dual to Rijndael from 240 to 9360. If this will have any impact on the security of Rijndael remains to be seen. Many properties of Rijndael, such as differential and linear probabilities, carry over to any of the duals, but other things can change. The designers of Rijndael stated in [2] that the constant $b$ in the affine transformation of the S-box was chosen so the S-box would have no fixed points. However, some of the duals have an S-box with four fixed points.

The idea of describing one of the duals of Rijndael completely in terms of $GF(2^2)$ did not pay off this time, but we hope it could serve as an inspiration to do more algebraic analysis of the AES.

# References

1. FIPS PUB 197. *Advanced Encryption Standard (AES)*, National Institute of Standards and Technology, U.S. Department of Commerce, November 2001.
   http://cscr.nist.gov/publications/fips/fips197/fips-197.pdf
2. J. Daemen, V. Rijmen. *AES Submission document on Rijndael, Version 2*, September 1999.
   http://csrc.nist.gov/CryptoToolkit/aes/rijndael/Rijndael.pdf
3. N. Ferguson, R. Schroeppel, D. Whiting. *A Simple Algebraic Representation of Rijndael.* Selected Areas in Cryptography 2001, LNCS 2259, pp. 103-111, 2001.
4. S. Murphy, M. Robshaw. *Essential Algebraic Structure within the AES.* CRYPTO 2002, LNCS 2442, pp. 1-16, 2002
5. E. Barkan, E. Biham. *In How Many Ways Can You Write Rijndael?.* ASIACRYPT 2002, LNCS 2501, pp. 160-175, 2002.
6. N. Courtois, J. Pieprzyk. *Cryptanalysis of Block Ciphers with Overdefined Systems of Equations.* ASIACRYPT 2002, LNCS 2501, pp. 267-287, 2002.
7. A. Biryukov, C. De Cannière, A. Braeken, B. Preneel. *A Toolbox for Cryptanalysis: Linear and Affine Equivalence Algorithms.* EUROCRYPT 2003, LNCS 2656, pp. 33-50, 2003.
8. J. Wolkerstorfer, E. Oswald, M. Lamberger. *An ASIC Implementation of the AES SBoxes.* CT-RSA 2002, LNCS 2271, pp. 67-78, 2002

# Representations and Rijndael Descriptions*

Vincent Rijmen and Elisabeth Oswald

IAIK, Graz University of Technology,
Inffeldgasse 16a, A-8010 Graz, Austria
{vincent.rijmen, elisabeth.oswald}@iaik.tugraz.at

**Abstract.** We discuss different descriptions of Rijndael and its components and how to find them. The fact that it is easy to find equivalent descriptions for the Rijndael transformations, has been used for two different goals. Firstly, to design implementations on a variety of platforms, both efficient and resistant against side channel analysis. Secondly, to analyze the security of the cipher We discuss these aspects, give examples, and present our views.

## 1  Introduction

In this paper, we give an overview of recent developments in the study of Rijndael security and efficient Rijndael implementations. Central to many of these developments is the technique of changing representations, and therefore we take this as the central theme of our treatment here.

When we look at what has been published about Rijndael in the last couple of years, we see that most authors restrict in their studies the possible changes of representation to the set of polynomial bases in a finite field, e.g. selection of a different base element or the selection of a different reduction polynomial. However, finite fields have a much richer structure, e.g. they can also be described as vector spaces over the ground field. It is our belief that exploration of the vector space representation can bring us to new insights in both security and efficient implementation of the Rijndael.

We start this paper by setting the framework to study different representations and the resulting equivalent descriptions for Rijndael. Afterwards, we present the overview of recent results and place them in our framework.

## 2  Change of Representation: An Old Mathematical Technique

It is well-known that the choice of representation influences the complexity of most problems related to algebra. One example with application in cryptography

---

* This research was supported financially by the A-SIT, Austria.

H. Dobbertin, V. Rijmen, A. Sowa (Eds.): AES 2004, LNCS 3373, pp. 148–158, 2005.

is given by elliptic curves. An arbitrary elliptic curve has a defining equation of the following form:

$$Ay^2 + Byx + Cy = Dx^3 + Ex^2 + Fx + G. \tag{1}$$

By choosing another representation, the defining equation can be transformed into the following form:

$$y^2 = x^3 + ax + b \ . \tag{2}$$

For all defining equations of the form (1), there is an equation of the form (2) defining an elliptic curve with the same mathematical properties, although both curves contain different points $(x, y)$.

A second example is the gate complexity of a circuit that implements the squaring operation in a finite field with characteristic two. If the elements of the field are represented by their coordinates with respect to a normal basis, then the squaring operation corresponds to a simple rotation of the coordinates. In other representations, the squaring operation corresponds to a more complicated linear transformation of the coordinates.

The problem we address in this paper, is exactly the opposite problem. When given a Boolean transformation with fixed 'points' $(x, S(x))$, we want to find a simple algebraic description for this Boolean transformation.

# 3  Boolean Transformations and Algebras

In order to improve understanding of the issues related to equivalent descriptions, it is important to clearly define the terminology. We make a distinction between two mathematical concepts that are often used as synonyms. These concepts are an *abstract element of an algebra* on the one hand, and the *representation of the element* on the other hand. We start with a definition for an algebra.

An *algebra* consists of one or more sets of elements and one or more *operations* between the elements. We will consider here algebras that contain only one set of elements, denoted by $\mathcal{A}$. Furthermore, we will assume that the cardinality of $\mathcal{A}$ equals $2^n$ for some integer value $n$.

An $m$-ary operation $b$ maps an input consisting of $m$ elements of $\mathcal{A}$ to an output, which is also in $\mathcal{A}$.

$$b : \mathcal{A}^m \to \mathcal{A} : (x_1, x_2, \ldots, x_m) \mapsto b(x_1, x_2, \ldots, x_m) = y \tag{3}$$

A 1-ary operation is also called an (algebraic) function.

A *Boolean vector* is a one-dimensional array of bits. By *Boolean transformation*, we mean a function $S$ that maps a Boolean vector to another Boolean vector. For sake of simplicity, we will assume that the input and output vectors have equal size.

$$S : Z_2^n \to Z_2^n : \mathrm{x} \mapsto \mathrm{y} = S(\mathrm{x}) \tag{4}$$

A *representation* $\rho$ maps the elements of $\mathcal{A}$ to $n$-bit Boolean vectors.

$$\rho : \mathcal{A} \to Z_2^n : x \to \rho(x) = \mathrm{x} \tag{5}$$

The inverse map of a representation is called a *labeling map*. A representation, or labeling, defines a map from the algebraic functions to the Boolean transformations:

$$R(b) = S_b \Leftrightarrow \forall x \in \mathcal{A} : \rho(b(x)) = S_b(\rho(x)) .$$ (6)

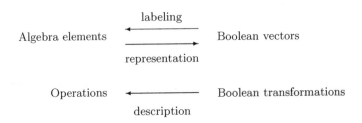

**Fig. 1.** Terminology w.r.t. algebra elements and Boolean vectors

### 3.1    Finding Descriptions

Finding an algebraic description for a Boolean transformation $S$ can be done in three steps.

**Initialization:** Decide on an algebra to be used.
**Labeling:** Define a labeling map from the Boolean vectors to elements of the algebra.
**Describing:** Compute the description of the Boolean transformation $S$.

Before describing the steps in some more detail, we briefly return to the previous example. Suppose we have a Boolean transformation $S$ that implements a rotation of the bits: $S(x) = x \lll 1$. Suppose further that we have chosen to use the finite field $GF(2^n)$ as the algebra to work in. Since $S$ is linear over this field, it can always be described with a linear polynomial $l(x)$:

$$l(x) = \sum_{i=0}^{n-1} a_i x^{2^i},$$ (7)

where the $a_i$ are some constants. If the Boolean vectors x are considered as the representation of the coordinates of the elements $x$ with respect to a normal basis, then the polynomial describing $S$ can be as simple as $l(x) = x^2$.

**Selecting an Algebra.** For the values of $n$ that are of practical importance, the number of algebras that can be defined, is very large: $(2^n)^{2^{2n}}$ different binary operations can be defined. However, the requirement to obtain a 'simple' description in practice limits the number of interesting algebras. The most natural choices are perhaps the vector space $(GF(2))^n$ or the field $GF(2^n)$. However, other algebras can be used as well, e.g. the algebra $< Z_{2^n+1} \backslash \{0\}, \times >$ as used in IDEA.

**Selecting a Labeling.** Once an algebra has been selected, the elements of the algebra have to be assigned to the Boolean vectors. We call this the *labeling* of the Boolean vectors. Let the labeling map be denoted by $m$:

$$m : (Z_2)^n \to \mathcal{A}. \tag{8}$$

The number of labeling maps equals $(2^n!)$. Note that this is more general than a change of basis in a finite field.

**Computing the Representation.** Once the algebra and the labeling map are defined, the input-output tuples of the transformation $S$ can be translated into tuples of the algebra:

$$(x, S(x)) \to (a, b) = (m(x), m(S(x))). \tag{9}$$

The functional description of $S$ in the new representation can be derived from the input-output tuples, using for instance the Lagrange interpolation formula, if it can be applied in the algebra selected.

# 4    Rijndael Descriptions

Equivalent descriptions can be investigated for any block cipher. Indeed, already in the 1980's, several results appeared about equivalent descriptions for the DES [6]. Afterwards, the topic seems to have died out in the field of symmetric cryptography. The selection of Rijndael to become the AES has triggered new research in this direction.

## 4.1    Number of Equivalent Descriptions

The design of Rijndael was made in the field GF(256). All design criteria were made and the selection of components was done with this field in mind. In order to be able to define the input-output behavior uniquely, one specific representation of the field elements had to be chosen. The choice was made to use a binary polynomial basis, with irreducible polynomial $p(t) = t^8 + t^4 + t^3 + t + 1$.

In [2], it was observed that 240 *equivalent ciphers* (i.e. alternative descriptions) can be generated by choosing one of the 30 irreducible binary polynomials of degree 8 and by choosing one of the 8 roots of this polynomial as generator.

However, there are many more alternative representations possible. The field GF(256) is isomorphic to the field $(GF(2))^8$, which is also an 8-dimensional vector space. In this vector space, there are

$$\prod_{i=0}^{7}(2^8 - 2^i) \approx 2^{62} \tag{10}$$

different bases. Each base leads to a different labeling of bytes and hence a different description of Rijndael. In [3], 2040 bases with a special property are derived.

The authors combine these 2040 bases with the 30 irreducible polynomials in order to define 61200 equivalent cipher descriptions.

Equivalent descriptions can also be constructed by defining an arbitrary bijective labeling map $m$. There are 256! different such labeling maps and at least as many different descriptions. Finally, observe that labeling maps don't have to be bijective. Also injective maps can be used (cf. infra), and hence there are infinitely many labeling maps.

### 4.2     Useful Representations

The vast majority of the 256! equivalent descriptions of Rijndael will not result in any new insights. Indeed, only the $2^{62}$ descriptions that are constructed following a change of basis in the vector space $(GF(2))^8$, have the property that 'addition' corresponds to binary exclusive-or. In all other descriptions, the specification of addition will require the use of tables without any apparent structure.

The transformations MixColumns and AddRoundKey can be described by very simple operations when the default representation is used. This is a second reason to look for new representations 'close' to the default representation.

## 5     Descriptions Assisting Implementations

Alternative descriptions facilitating implementations are used mostly on constrained platforms: hardware and small processors. In environments with little constraints, the default description of Rijndael seems to be as good as any other one.

The addition of side-channel attack countermeasures usually decreases the performance and/or increases the cost of an implementation. Hence, the techniques developed to improve the performance of ordinary implementations in constrained environments, are usually also of use in side-channel attack resisting hardware.

### 5.1     Hardware Efficient Descriptions

As explained before, the use of alternative finite field representations in order to reduce the gate count of a circuit is a well-established technique. Several alternative representations have been proposed in the cryptographic literature, mainly in order to improve the implementation of the SubBytes transformation.

The first type of alternative representation is to label bytes as polynomials of degree smaller than 2, with coefficients in GF(16):

$$m : Z_2^8 \rightarrow GF(16)[t]/(t^2 + At + B) : x \mapsto m(x) = at + b, \tag{11}$$

with $a = m_1(x), b = m_2(x)$. The maps $m_1, m_2$ have to satisfy some conditions and the coefficients $A, B$ are chosen such that the polynomial $t^2 + At + B$ is irreducible. Then, the transformation SubBytes can be described by one nonlinear formula of the form

$$(at + b)^{-1} = (b^2 B + baA + a^2)^{-1}(bt + a + bA), \qquad (12)$$

combined with linear and affine operations, which depend on the choice for $A, B$ and the details of the maps $m_1(x), m_2(x)$. Descriptions of this type have been proposed in [12, 13, 14, 16].

The alternative representations mainly improve the gate complexity of the SubBytes step, while the complexity of the other steps remains the same, or deteriorates slightly. This results in different approaches. In the first approach, the SubBytes is implemented in the representation that is best for that step, and the other steps are implemented using another representation. This approach necessitates changes of representation in between steps [14, 16].

In the second approach, frequent changes of representation are avoided by adopting a 'compromise' representation, which improves the complexity of Sub-Bytes, and doesn't increase the complexity of other steps too much [13].

In the third approach, an alternativ representation, which is best for the SubBytes step, is combined with an equivalent AES [17].

## 5.2    Representations Assisting SCA Countermeasures

Side-channel attacks are used to extract secret key material from real systems. It has been observed that computing hardware and software often leak information about secret keys used in cryptographic algorithms. This leakage comes from variations in execution time, power consumption, radiation, etc.

Many proposals for hardware designs that resist side-channel attacks, are based on *masking* techniques: the sensitive values are never manipulated directly, but only in blinded, or masked, form. The mask is a random value, which needs to be processed separately. Such a masking scheme can also be described as a secret sharing scheme, where the masked and the masked value are two *shares*. Hence, a masking scheme by itself can already be seen as an expanding alternative representation.

For linear operations, it is well-known what masking schemes to use and how to implement them. For the non-linear operation of Rijndael, there are several proposals.

## 5.3    Additive Split

In order to implement a linear operation $L(x)$, the input $x$ is represented by a tuple $(p, q)$ with $p + q = x$. We call this the *additive split* of sensitive variables. In order to compute the tuple corresponding to $L(x)$, the linear operation is performed separately on each share. Indeed, we have that if $x = p + q$, then $L(x) = L(p) + L(q)$, and hence $L(x)$ is represented by $(L(p), L(q))$.

The addition of two sensitive variables can also be protected using the additive split. The result can be computed in a secure way by simply adding the co-ordinates of the corresponding tuples: if $x$ is represented by $(p, q)$ and $y$ by $(r, s)$, then $x + y$ can be represented by $(v, w) = (p + r, q + s)$ or $(v, w) = (p + s, q + r)$. In both representations, $v$ and $w$ are completely uncorrelated to the values of $x$ and $y$.

The SubBytes step in the Rijndael round transformation consists of an affine operation and the multiplicative inverse map, or, more accurately, the power function map $x^{254}$. Protecting non-linear maps by means of an additive split is not a straightforward process.

In order to implement this map, two functions $f, g$ are required, such that $x^{254} = f(p) + g(q)$ (where $x = p + q$). In order to have security against first-order side-channel attacks, the implementation of maps $f(p)$ and $g(q)$ should not produce intermediate results which correlate with $p + q$. It remains an open problem whether such maps can be defined.

As an alternative solution, other types of split have been proposed in the literature and we describe them in Section 5.4. In Section 5.5, we describe a recently developed method. By using a special representation of the field elements, it becomes possible to protect also the power function by means of an additive split.

## 5.4     Multiplicative Split

A multiplicative split was proposed in [1]. A byte $x$ is mapped to $(p, q)$ with $x = pq^{254}$. It can be seen that the tuple corresponding to $x^{254}$ can be computed as $(p^{254}, q^{254}) = (q, p)$. Hence this representation would allow to implement the power function with a simple swap of registers. Alas, it seems from [1] that the requirement to change the representation from additive split to multiplicative split and vice versa, makes it necessary to compute the power functions of the two shares explicitly.

A more important disadvantage is the so-called zero multiplication problem, which refers to the fact that a multiplicative split fails to hide the zero value:

$$x = 0 \Leftrightarrow (p = 0 \text{ or } q = 0). \tag{13}$$

One approach to fix this problem is to introduce a second injective map, that maps the elements of GF(256) to a larger ring containing zero divisors [9].

## 5.5     Additive Split in Tower Fields

This technique has been described in [11]. It builds on the techniques explained in [16]. During all operations, the variables are protected by means of an additive split. In [4], an alternative method was developed, based on the same principles.

In the tower field representation, bytes are labeled as polynomials of degree smaller than 2. The two coefficients of the polynomials are elements of GF(16). For some of the computations, these coefficients in turn are labeled as polynomials of degree 2, with coefficients in GF(4).

$$m : Z_2^8 \to GF(16)[t]/(t^2 + At + B) : x \mapsto m(x) = at + b, \tag{14}$$
$$m' : GF(16) \to GF(4)[u]/(u^2 + Cu + D) : y \mapsto m'(y) = cu + d, \tag{15}$$

with $a = m_1(x), b = m_2(x), c = m'_1(y), d = m'_2(y)$. The maps $m_1, m_2, m'_1, m'_2$ have to satisfy some conditions and the coefficients $A, B, C, D$ are chosen such that the polynomials $t^2 + At + B$ and $u^2 + Cu + D$ are irreducible over $GF(16)$, respectively $GF(4)$, and lead to efficient arithmetic. Note that the labeling maps are all linear and hence they can be protected as explained in Section 5.3. The multiplicative inverse map can be described in two steps, which are explained below.

**Step 1: GF(16).** Firstly, (12) is used to describe the map using only the following operations: addition, multiplication and taking the multiplicative inverse. All these operations are in $GF(16)$. The implementation of the multiplicative inverse is done in Step 2. The additions in $GF(16)$ are protected as explained in Section 5.3.

The multiplication of two sensitive variables $a$ and $b$, that are represented by $(p, q)$ and $(r, s)$, is implemented by multiplying the coordinates $p$ and $r$, and adding so-called *correction terms* $c_i$. The correction terms are defined in such a way that they can be computed without producing intermediate results that correlate to $a$, $b$, $ab$, $a + b$, $a^2$, $b^2$, or any other value that could leak information about the sensitive variables to the attacker. The result is a representation of the following form:

$$(v, w) = (pq + \sum_i c_i, q). \tag{16}$$

**Step 2: GF(4).** The second step is similar to the first step, but operating on smaller fields. A formula very similar to (12) describes how the multiplicative inverse in $GF(16)$ can be computed using operation in $GF(4)$ only.

Addition and multiplication in $GF(4)$ are implemented and protected as described for Step 1. For the implementation of the multiplicative inverse, the following fact is used.

For all $x \in GF(4)$, it holds that $x^{-1} = x^2$, hence taking the multiplicative inverse is a linear operation, that can be protected by means of an additive split, as described in Section 5.3.

# 6   Representations Assisting Cryptanalysis

Several alternative descriptions have been derived, showing that more elegant, more structured and more *simple* sets of equations defining Rijndael can be constructed. Although several of them seem a promising start for an attack, no breakthrough has been demonstrated yet.

## 6.1   BES

Murphy and Robshaw [10] define the block cipher BES, which operates on data blocks of 128 bytes instead of bits. According to Murphy and Robshaw, the algebraic structure of BES is even more elegant and simple than that of Rijndael. Furthermore, Rijndael can be *embedded* into BES. There is a map $\phi$ such that:

$$\text{Rijndael}(x) = \phi^{-1}\left(\text{BES}\left(\phi(x)\right)\right). \tag{17}$$

The map $\phi$ can also be seen as an injective labeling of the inputs of Rijndael.

Murphy and Robshaw proceed with some observations on the properties of BES. However, these properties of BES do not apply to Rijndael.

## 6.2    Redundant S-Boxes

Any $8 \times 8$-bit S-box can be considered as a composition of 8 Boolean functions sharing the same 8 input bits. J. Fuller and W. Millan observed that the S-box of Rijndael can be described using one Boolean function only [8]. The 8 Boolean functions can be described as

$$f_i(x_1, \ldots, x_8) = f(g_i(x_1, \ldots, x_8)) + c_i, \qquad i = 1, \ldots, 8, \qquad (18)$$

where the function $f$ is the only nonlinear function, the $g_i$ are affine functions and the $c_i$ are constants.

## 6.3    Continued Fractions

Ferguson, Schroeppel and Whiting [7] derive a closed formula for Rijndael that can be seen as a generalization of continued fractions. Any byte of the intermediate result after 5 rounds can be expressed as follows.

$$x = K + \sum \cfrac{C_1}{K^* + \sum \cfrac{C_2}{K^* + \sum \cfrac{C_3}{K^* + \sum \cfrac{C_4}{K^* + \sum \cfrac{C_5}{K^* + p^*_*}}}}} \qquad (19)$$

Here every $K$ is some expanded key byte, each $C_i$ is a known constant and each $*$ is a known exponent or subscript, but these values depend on the summation variables that enclose the symbol.

A fully expanded version of (19) has $2^{25}$ terms. It is currently unknown what a practical algorithm to solve this type of equations would look like.

## 6.4    XL and XSL Methods

XL and XSL are new methods to solve nonlinear algebraic equations [5, 15]. The effectiveness and efficiency of these methods remain a topic of debate. It seems plausible that the complexity of the methods is influenced by the description that is chosen. For instance, the authors of [10] claim that using their representation, the complexity of the XSL method decreases significantly. More details about the methods can be found elsewhere in these proceedings.

## 7    Conclusions and Perspective on the Future

Compared with other symmetric ciphers, the design of Rijndael shows a remarkable level of mathematical abstraction. The rich structure, and in particular the ease with which equivalent descriptions can be constructed, on the one hand has caused worries with some people fearing for its long-term security. On the other hand, the presence mathematical structure has certainly greatly facilitated the development of efficient implementations on constrained environments and environments where protection measures against side-channel attacks are required.

We expect that the continuation and generalization of this research will lead to better insights in the security of Rijndael and even more efficient implementations. Furthermore, we expect that many results will be applicable to other ciphers as well, leading to an increased understanding about the process of designing secure symmetric primitives.

# References

1. Mehdi-Laurant Akkar and Christophe Giraud, "An implementation of DES and AES secure against some attacks," *CHES 2001*, LNCS 2162, Springer-Verlag, 2001, pp. 309–318.
2. Elad Barkan and Eli Biham, "In how many ways can you write Rijndael?", Advances in Cryptology — Asiacrypt 2002, LNCS 2051, Springer-Verlag, 2002, pp. 160–175.
3. Alex Biryukov, Christophe De Cannière, An Braeken and Bart Preneel, "A toolbox for cryptanalysis: linear and affine equivalence algorithms," *Advances in Cryptology — Eurocrypt 2003*, LNCS 2656, Springer-Verlag, 2003, pp. 33–50.
4. Johannes Blömer, Guajardo Merchan and Volker Krummel, "Provably secure masking of AES", *Selected Areas in Cryptography - SAC'04*, LNCS, Springer-Verlag, to appear.
5. Nicolas T. Courtois and Josef Pieprzyk, "Cryptanalysis of block ciphers with overdefined systems of equations," *Advances in Cryptology — Asiacrypt '02*, LNCS 2501, Springer-Verlag, 2003, pp. 267–287.
6. Marc Davio, Yvo Desmedt, Marc Fosséprez, René Govaerts, Jan Hulsbosch, Patrik Neutjens, Philippe Piret, Jean-Jacques Quisquater, Joos Vandewalle, and Pascal Wouters, "Analytical characteristics of the DES," *Advances in Cryptology — Crypto '83*, Plenum Press, 1984, pp. 171–202.
7. Niels Ferguson, Richard Schroeppel, and Doug Whiting, "A simple algebraic representation of Rijndael," *Selected Areas in Cryptography  SAC01*, LNCS 2259, Springer-Verlag, 2001, pp. 103-111.
8. Joanne Fuller and William Millan, "On linear redundancy in S-boxes," *Fast Software Encryption '03*, LNCS 2887, Springer-Verlag, 2003, pp. 74–86.
9. Jovan Dj. Golic and Christophe Tymen, "Multiplicative masking and power analysis of AES," *CHES 2002*, LNCS 2535, Springer-Verlag, 2003, pp. 198–212.
10. Sean Murphy and Matthew J.B. Robshaw, "Essential algebraic structure within the AES", *Advances in Cryptology — Crypto 2002*, LNCS 2442, Springer-Verlag, 2002, pp. 17–38.
11. Elisabeth Oswald, Stefan Mangard and Norbert Pramstaller, "Secure and efficient masking of the AES: a mission impossible?," *Technical report IAIK-TR 2003/11/1*, available from http://eprint.iacr.org/2004/134.pdf.
12. Vincent Rijmen, "Efficient implementation of the Rijndael S-box," available from http://www.esat.kuleuven.ac.be/~rijmen/rijndael/sbox.pdf, 2000.

13. Atri Rudra, Pradeep K. Dubey, Charanjit S. Jutla, Vijay Kumar, Josyula R. Rao, and Pankaj Rohatgi, "Efficient Rijndael encryption implementation with composite field arithmetic," *CHES 2001*, LNCS 2162, Springer-Verlag, 2001, pp. 171–184.
14. Akashi Satoh, Sumio Morioka, Kohji Takano, and Seiji Munetoh, "A compact Rijndael hardware architecture with S-box optimization," *Advances in Cryptology, Asiacrypt 2001*, LNCS 2248, Springer-Verlag, 2001, pp. 239–254.
15. Adi Shamir, Jacques Patarin, Nicolas Courtois and Alexander Klimov, "Efficient Algorithms for solving Overdefined Systems of Multivariate Polynomial Equations", *Advances in Cryptology — Eurocrypt 2000*, LNCS 1807, Springer-Verlag, 2000, pp. 392–407.
16. Johannes Wolkerstorfer, Elisabeth Oswald, and Mario Lamberger, "An ASIC implementation of the AES S-boxes," *Topics in Cryptology — CT-RSA 2002*, LNCS 2271, Springer-Verlag, 2002, pp. 67–78.
17. Shee-You Wu, Shih-Chuan Lu, and Chi Sung Laih, "Design of AES Based on Dual Cipher and Composite Field," *Topics in Cryptology — CT-RSA 2004*, LNCS 2964, Springer-Verlag, 2004, pp. 25–38

# Linearity of the AES Key Schedule

Frederik Armknecht[*] and Stefan Lucks

Theoretische Informatik,
Universität Mannheim,
68131 Mannheim Germany
armknecht@th.informatik.uni-mannheim.de
lucks@th.informatik.uni-mannheim.de

**Abstract.** The AES key schedule can *almost* be described as collection of 32 linear feedback shift registers LFSRs, working in parallel. This implies that for *related keys*, i.e., pairs of unknown keys with known differences, one can in part predict the differences of the individual round keys. Such a property has been used (but not explained in detail) by Ferguson et al. [3] for a related key attack on a 9-round variant of the AES (with 256-bit keys). In the current paper, we study the propagation of (known) key differences in the key schedule for all three key sizes of the AES.

## 1   Introduction

Recall the key schedule of the AES, e.g. for 128-bit keys. Denote the inital *cipher key* $(K_0, K_1, K_2, K_3) \in \left( \{0,1\}^{32} \right)^4$. This is the first round key, as well. The next round key is $(K_4, K_5, K_6, K_7)$, the next one is $(K_8, K_9, K_{10}, K_{11})$, ... The 32-bit values $K_i$ are generated by the following *key schedule* algorithm:

> If $(i \bmod 4) = 0$
> then $K_i := K_{i-4} \oplus f(K_{i-1}) \oplus \mathrm{const}(i)$
> else $K_i := K_{i-4} \oplus K_{i-1}$,

where $f : \{0,1\}^{32} \rightarrow \{0,1\}^{32}$ is a nonlinear function and $\mathrm{const}(i)$ are some round-dependent constants. The nonlinear function $f$ allows reasonably efficient implementations, and thus the key schedule itself is highly efficient. The key schedules for 192-bit and 256-bit keys are defined similarly.

Consider two unknown cipher keys $(K_0, K_1, K_2, K_3)$ and $(\tilde{K}_0, \tilde{K}_1, \tilde{K}_2, \tilde{K}_3)$ with known differences $\delta_i = K_i \oplus \tilde{K}_i$, $i = 0, 1, 2, 3$. The key schedule allows us to describe linear realationships of the form

$$\bigoplus_{i=0}^{43} c_i \cdot (K_i \oplus \tilde{K}_i) = \delta, \quad c_i \in \{0,1\}$$

---

[*] Supported by grant 620307 of the DFG (German Research Foundation).

H. Dobbertin, V. Rijmen, A. Sowa (Eds.): AES 2004, LNCS 3373, pp. 159–169, 2005.

with known $\delta$. In principle, such relationships could be used to mount *related-key attacks* against the AES – and in fact, one such related-key attack for the AES variant with 256-bit keys has been previously published [3].

The current paper investigates the existence of such linear relationships in the AES key schedule(s). It turns out that for none of the defined key sizes, any such relationship exists which covers the entire key-schedule (i.e., which involves values from the first round key *and* values from the last round key, but no values from round keys in between).

## 2    Definitions and Motivation

There exist three different versions of the AES:

| $N_k$ | Key size $32 \cdot N_k$ | Number of rounds $N_r$ |
|---|---|---|
| 4 | 128 bit | 10 |
| 6 | 192 bit | 12 |
| 8 | 256 bit | 14 |

Before encryption, the secret key $\mathcal{K}$ of size $32 \cdot N_k$ is expanded to a key $K$ of size $128 \cdot (N_r + 1)$. Following the description given in [2] we divide the expanded key $K$ into $4 \cdot (N_r + 1)$ parts $K_0, \ldots, K_{4N_r+3} \in \{0,1\}^{32}$. In the following, we will treat the $K_i$ as vectors of the vector space $\mathbb{F}_2^{32}$ and denote by $\oplus$ the corresponding addition of two vectors.[1] The first $N_k$ vectors are exactly the secret key $\mathcal{K}$ and the rest is defined by the key schedule.

To increase the resistance against related-key attacks the key schedule was designed such that the descriptions of the columns $K_i$, $i \geq N_k$, are linearly independent of $\mathcal{K}$. In fact, given two unknown keys $\mathcal{K}$ and $\tilde{\mathcal{K}}$ with known difference

$$\mathcal{K} \oplus \tilde{\mathcal{K}} = (K_0 \oplus \tilde{K}_0, \ldots, K_{N_k-1} \oplus \tilde{K}_{N_k-1}) =: (\delta_0, \ldots, \delta_{N_k-1}) =: \delta$$

it should be infeasible to say anything about the differences $K_i \oplus \tilde{K}_i$ for $i \geq N_k$. Surprisingly, it is possible to find (many) linear combinations of the following kind:

$$\bigoplus_{i=0}^{4N_r+3} c_i \cdot K_i = 0, \quad c_i \in \mathbb{F}_2 \tag{1}$$

This implies that the following equation is true

$$\bigoplus_{i=0}^{N_k-1} c_i \cdot \delta_i = \bigoplus_{i=0}^{N_k-1} c_i \cdot (K_i \oplus \tilde{K}_i) = \bigoplus_{i=N_k}^{4N_r+3} c_i \cdot (K_i \oplus \tilde{K}_i)$$

For example the following equation holds for the 128-bit variant:

$$K_4 \oplus \tilde{K}_4 \oplus K_5 \oplus \tilde{K}_5 = \delta_1$$

In the following section, we develop the general theory and provide a basis of all valid linear combinations of $K_0, \ldots, K_{4N_r+3}$ for all three AES variants.

---

[1] In fact, this is simply the componentwise XOR of the 32 bits of the two vectors.

## 3    Linearity of the AES Key Schedule

In the following we examine the key schedules of the three AES variants with 128-bit, 192-bit resp. 256-bit key lengths. We will see that in all cases many equations of the type (1) exist.

### 3.1    AES with 128 Bit Key Length

Let $\mathcal{K} = (\mathcal{K}_0, \mathcal{K}_1, \mathcal{K}_2, \mathcal{K}_3) \in \{0,1\}^{128}$ be the secret key with $\mathcal{K}_i \in \{0,1\}^{32}$. Then the expanded key $K = (K_0, \ldots, K_{43})$ is defined by the following key schedule:

$$
\begin{aligned}
K_i &:= \mathcal{K}_i & &, 0 \leq i < 4 \\
K_i &:= K_{i-4} \oplus f_i(K_{i-1}) & &, 4 \leq i \leq 43,\ i \bmod 4 = 0 \\
K_i &:= K_{i-4} \oplus K_{i-1} & &, 4 \leq i \leq 43,\ i \bmod 4 \neq 0
\end{aligned}
$$

$f_i(x)$ is the permutation $f(x) \oplus const(i)$ mentionend in section 1; but we will see that the exact definition of $f_i$ does not matter for our observations. To motivate the theory, we have a look at the definition of $K_4, \ldots, K_{11}$:

$$
\begin{aligned}
K_4 &= K_0 \oplus f_4(K_3) \\
K_5 &= K_1 \oplus K_4 = K_0 \oplus K_1 \oplus f_4(K_3) \\
K_6 &= K_2 \oplus K_5 = K_0 \oplus K_1 \oplus K_2 \oplus f_4(K_3) \\
K_7 &= K_3 \oplus K_6 = K_0 \oplus K_1 \oplus K_2 \oplus K_3 \oplus f_4(K_3) \\
K_8 &= K_4 \oplus f_8(K_7) = K_0 \oplus f_4(K_3) \oplus f_8(K_7) \\
K_9 &= K_8 \oplus K_5 = K_1 \oplus f_8(K_7) \\
K_{10} &= K_9 \oplus K_6 = K_0 \oplus K_2 \oplus f_4(K_3) \oplus f_8(K_7) \\
K_{11} &= K_{10} \oplus K_7 = K_1 \oplus K_3 \oplus f_8(K_7)
\end{aligned}
$$

We observe that each of the vectors $K_0, \ldots, K_{11}$ can be expressed by a linear combination of $K_0, K_1, K_2, K_3, f_4(K_3), f_8(K_7)$. This can be easily generalized: each of the vectors $K_0, \ldots, K_{43}$ can be written as a linear combination of elements of the set

$$
B := \{\underbrace{K_0, K_1, K_2, K_3}_{=\mathcal{K}}, f_4(K_3), \ldots, f_{40}(K_{39})\}.
$$

This complies with the following matrix-vector-product where $M$ is a binary matrix of size $44 \times 14$:

$$
M \cdot \begin{pmatrix} K_0 \\ \vdots \\ K_3 \\ f_4(K_3) \\ \vdots \\ f_{40}(K_{39}) \end{pmatrix} = \begin{pmatrix} K_0 \\ \vdots \\ K_{43} \end{pmatrix} \tag{2}
$$

| | $K_0$ | $K_1$ | $K_2$ | $K_3$ | $f_4(K_3)$ | $\dots$ | $f_{40}(K_{39})$ |
|---|---|---|---|---|---|---|---|
| $K_0$ | 1 | 0 | 0 | 0 | 0 | 0000 0 000 | 0 |
| | 0 | 1 | 0 | 0 | 0 | 0000 0 000 | 0 |
| | 0 | 0 | 1 | 0 | 0 | 0000 0 000 | 0 |
| | 0 | 0 | 0 | 1 | 0 | 0000 0 000 | 0 |
| $K_4$ | 1 | 0 | 0 | 0 | 1 | 0000 0 000 | 0 |
| | 1 | 1 | 0 | 0 | 1 | 0000 0 000 | 0 |
| | 1 | 1 | 1 | 0 | 1 | 0000 0 000 | 0 |
| | 1 | 1 | 1 | 1 | 1 | 0000 0 000 | 0 |
| $K_8$ | 1 | 0 | 0 | 0 | 1 | 1000 0 000 | 0 |
| | 0 | 1 | 0 | 0 | 0 | 1000 0 000 | 0 |
| | 1 | 0 | 1 | 0 | 1 | 1000 0 000 | 0 |
| | 0 | 1 | 0 | 1 | 0 | 1000 0 000 | 0 |
| $K_{12}$ | 1 | 0 | 0 | 0 | 1 | 1100 0 000 | 0 |
| | 1 | 1 | 0 | 0 | 1 | 0100 0 000 | 0 |
| | 0 | 1 | 1 | 0 | 0 | 1100 0 000 | 0 |
| | 0 | 0 | 1 | 1 | 0 | 0100 0 000 | 0 |
| $K_{16}$ | 1 | 0 | 0 | 0 | 1 | 1110 0 000 | 0 |
| | 0 | 1 | 0 | 0 | 0 | 1010 0 000 | 0 |
| | 0 | 0 | 1 | 0 | 0 | 0110 0 000 | 0 |
| | 0 | 0 | 0 | 1 | 0 | 0010 0 000 | 0 |
| $K_{20}$ | 1 | 0 | 0 | 0 | 1 | 1111 0 000 | 0 |
| | 1 | 1 | 0 | 0 | 1 | 0101 0 000 | 0 |
| | 1 | 1 | 1 | 0 | 1 | 0011 0 000 | 0 |
| | 1 | 1 | 1 | 1 | 1 | 0001 0 000 | 0 |
| $K_{24}$ | 1 | 0 | 0 | 0 | 1 | 1111 1 000 | 0 |
| | 0 | 1 | 0 | 0 | 0 | 1010 1 000 | 0 |
| | 1 | 0 | 1 | 0 | 1 | 1001 1 000 | 0 |
| | 0 | 1 | 0 | 1 | 0 | 1000 1 000 | 0 |
| $K_{28}$ | 1 | 0 | 0 | 0 | 1 | 1111 1 100 | 0 |
| | 1 | 1 | 0 | 0 | 1 | 0101 0 100 | 0 |
| | 0 | 1 | 1 | 0 | 0 | 1100 1 100 | 0 |
| | 0 | 0 | 1 | 1 | 0 | 0100 0 100 | 0 |
| $K_{32}$ | 1 | 0 | 0 | 0 | 1 | 1111 1 110 | 0 |
| | 0 | 1 | 0 | 0 | 0 | 1010 1 010 | 0 |
| | 0 | 0 | 1 | 0 | 0 | 0110 0 110 | 0 |
| | 0 | 0 | 0 | 1 | 0 | 0010 0 010 | 0 |
| $K_{36}$ | 1 | 0 | 0 | 0 | 1 | 1111 1 111 | 0 |
| | 1 | 1 | 0 | 0 | 1 | 0101 0 101 | 0 |
| | 1 | 1 | 1 | 0 | 1 | 0011 0 011 | 0 |
| | 1 | 1 | 1 | 1 | 1 | 0001 0 001 | 0 |
| $K_{40}$ | 1 | 0 | 0 | 0 | 1 | 1111 1 111 | 1 |
| | 0 | 1 | 0 | 0 | 0 | 1010 1 010 | 1 |
| | 1 | 0 | 1 | 0 | 1 | 1001 1 001 | 1 |
| | 0 | 1 | 0 | 1 | 0 | 1000 1 000 | 1 |

**Fig. 1.** The linear expressions of $K_0, \dots, K_{43}$ for the 128-bit variant

Figure 1 in section 4 displays the linear expression for $K_0, \ldots, K_{43}$.

As the rank of $M$ is 14, one can construct 30 linearly independent vectors $C^{(1)}, \ldots, C^{(30)} \in \mathbb{F}_2^{32}$ such that

$$\left(C^{(i)}\right)^t \cdot M = \mathbf{0}, \quad i = 1, \ldots, 30.$$

In fact, $C^{(1)}, \ldots, C^{(30)}$ is a basis of the nullspace of $M^t$. Together with (2) this implies for $C^{(i)} = (c_0^{(i)}, \ldots, c_{43}^{(i)})^t$ the following equation:

$$0 = \left(C^{(i)}\right)^t \cdot \begin{pmatrix} K_0 \\ \vdots \\ K_{43} \end{pmatrix} = \bigoplus_{j=0}^{43} c_j^{(i)} \cdot K_j.$$

This is exactly an equation as displayed in (1). A possible choice of $C^{(1)}, \ldots, C^{(30)}$ is given in Figure 2 in section 4. We checked that no non-trivial linear relations between the $K_0, K_1, K_2, K_3 \, (= \mathcal{K})$ and the key vectors $K_{40}, K_{41}, K_{42}, K_{43}$ of the last round exist.

Assume we try to find expressions $\bigoplus_{i=0}^{43} c_i \cdot K_i = 0$ with at least one non-zero coefficient $c_{40}, \ldots, c_{43}$ and as many zero coefficients $c_{39}, c_{38}, \ldots$ as possible.[2] One such example is

$$K_2 \oplus K_3 \oplus K_8 \oplus K_{12} \oplus K_{24} \oplus K_{28} \oplus K_{40} \oplus K_{41} \oplus K_{42} \oplus K_{43}.$$

As one of the anonymous referees pointed out, this is optimal. It is straightforward to verify this using Figure 2.

Similarly, Figure 2 can be used to solve the open problem posed by Nicolas Courtois [1] whether $\bigoplus_{i=1}^{43} K_i$ is equal to zero. Figure 2 desribes a base for all valid linear combinations of the $K_i$. As it turns out, $\bigoplus_{i=1}^{43} K_i$ is not within its linear span.

## 3.2 AES with 192 Bit Key Length

We denote again by $\mathcal{K} = (\mathcal{K}_0, \mathcal{K}_1, \mathcal{K}_2, \mathcal{K}_3, \mathcal{K}_4, \mathcal{K}_5) \in \{0,1\}^{192}$ the secret key with $\mathcal{K}_i \in \mathbb{F}_2^{32}$. The key schedule is very similar to the 128 bit variant described in 3.1:

$$\begin{aligned} K_i &:= \mathcal{K}_i && ,0 \leq i < 6 \\ K_i &:= K_{i-6} \oplus f_i(K_{i-1}) && ,6 \leq i \leq 51, \; i \bmod 6 = 0 \\ K_i &:= K_{i-6} \oplus K_{i-1} && ,6 \leq i \leq 51, \; i \bmod 6 \neq 0 \end{aligned}$$

Again, the exact definition of $f_i$ is of no importance and is therefore omitted here.

---

[2] This means to express a linear relationship of the last round key $(K_{40}, K_{41}, K_{42}, K_{43})$ by earlier round keys - the earlier, the better.

| K₀   K₃ | | K₈ | | K₁₆ | | K₂₄ | | K₃₂ | | K₄₀ |
|---|---|---|---|---|---|---|---|---|---|---|
| 0 1 0 0 | 1 0 0 0 | 1 0 0 0 | 1 0 0 0 | 1 0 0 0 | 1 0 0 0 | 1 0 0 0 | 1 1 0 0 | 0 0 0 0 | 0 0 0 0 | 0 0 0 |
| 0 0 0 1 | 0 0 0 0 | 0 0 0 0 | 1 0 0 0 | 1 0 0 0 | 0 0 0 0 | 0 0 0 0 | 1 0 0 0 | 1 0 0 1 | 0 0 0 0 | 0 0 0 |
| 0 1 0 0 | 1 0 0 0 | 1 0 0 0 | 1 0 0 0 | 1 0 0 0 | 1 0 0 0 | 1 0 0 0 | 1 0 0 0 | 1 0 0 0 | 1 1 0 0 | 0 0 0 |
| 0 1 1 0 | 1 0 0 0 | 0 0 0 0 | 1 0 0 0 | 0 0 0 0 | 1 0 0 0 | 0 0 0 0 | 0 0 0 0 | 1 0 0 0 | 1 0 1 0 | 0 0 0 |
| 0 1 1 1 | 1 0 0 0 | 0 0 0 0 | 0 0 0 0 | 1 0 0 0 | 1 0 0 0 | 0 0 0 0 | 0 0 0 0 | 1 0 0 0 | 1 0 0 1 | 0 0 0 |
| 0 1 1 0 | 1 0 0 0 | 0 0 0 0 | 1 0 0 0 | 0 0 0 0 | 1 0 0 0 | 0 0 0 0 | 1 0 1 0 | 0 0 0 0 | 0 0 0 0 | 0 0 0 |
| 0 1 0 0 | 1 0 0 0 | 1 0 0 0 | 1 0 0 0 | 1 0 0 0 | 1 0 0 0 | 1 0 0 0 | 1 0 0 0 | 1 0 0 0 | 1 0 0 0 | 1 1 0 0 |
| 0 0 1 0 | 0 0 0 0 | 1 0 0 0 | 0 0 0 0 | 1 0 0 0 | 0 0 0 0 | 1 0 0 0 | 0 0 0 0 | 1 0 0 0 | 0 0 0 0 | 1 0 1 0 |
| 0 1 0 1 | 1 0 0 0 | 1 0 0 0 | 0 0 0 0 | 0 0 0 0 | 1 0 0 0 | 1 0 0 0 | 0 0 0 0 | 0 0 0 0 | 1 0 0 0 | 1 0 0 1 |
| 0 0 1 1 | 0 0 0 0 | 1 0 0 0 | 1 0 0 0 | 0 0 0 0 | 0 0 0 0 | 1 0 0 0 | 1 0 0 1 | 0 0 0 0 | 0 0 0 0 | 0 0 0 |
| 0 1 0 0 | 1 0 0 0 | 1 0 0 0 | 1 0 0 0 | 1 0 0 0 | 1 0 0 0 | 1 0 0 0 | 1 0 0 0 | 1 1 0 0 | 0 0 0 0 | 0 0 0 |
| 0 0 1 0 | 0 0 0 0 | 1 0 0 0 | 0 0 0 0 | 1 0 0 0 | 0 0 0 0 | 1 0 0 0 | 0 0 0 0 | 1 0 1 0 | 0 0 0 0 | 0 0 0 |
| 0 1 0 0 | 1 1 0 0 | 0 0 0 0 | 0 0 0 0 | 0 0 0 0 | 0 0 0 0 | 0 0 0 0 | 0 0 0 0 | 0 0 0 0 | 0 0 0 0 | 0 0 0 |
| 0 1 1 0 | 1 0 1 0 | 0 0 0 0 | 0 0 0 0 | 0 0 0 0 | 0 0 0 0 | 0 0 0 0 | 0 0 0 0 | 0 0 0 0 | 0 0 0 0 | 0 0 0 |
| 0 1 1 1 | 1 0 0 1 | 0 0 0 0 | 0 0 0 0 | 0 0 0 0 | 0 0 0 0 | 0 0 0 0 | 0 0 0 0 | 0 0 0 0 | 0 0 0 0 | 0 0 0 |
| 0 1 0 0 | 1 0 0 0 | 1 1 0 0 | 0 0 0 0 | 0 0 0 0 | 0 0 0 0 | 0 0 0 0 | 0 0 0 0 | 0 0 0 0 | 0 0 0 0 | 0 0 0 |
| 0 0 1 0 | 0 0 0 0 | 1 0 1 0 | 0 0 0 0 | 0 0 0 0 | 0 0 0 0 | 0 0 0 0 | 0 0 0 0 | 0 0 0 0 | 0 0 0 0 | 0 0 0 |
| 0 1 0 1 | 1 0 0 0 | 1 0 0 1 | 0 0 0 0 | 0 0 0 0 | 0 0 0 0 | 0 0 0 0 | 0 0 0 0 | 0 0 0 0 | 0 0 0 0 | 0 0 0 |
| 0 1 0 0 | 1 0 0 0 | 1 0 0 0 | 1 1 0 0 | 0 0 0 0 | 0 0 0 0 | 0 0 0 0 | 0 0 0 0 | 0 0 0 0 | 0 0 0 0 | 0 0 0 |
| 0 1 1 0 | 1 0 0 0 | 0 0 0 0 | 1 0 1 0 | 0 0 0 0 | 0 0 0 0 | 0 0 0 0 | 0 0 0 0 | 0 0 0 0 | 0 0 0 0 | 0 0 0 |
| 0 0 1 1 | 0 0 0 0 | 1 0 0 0 | 1 0 0 1 | 0 0 0 0 | 0 0 0 0 | 0 0 0 0 | 0 0 0 0 | 0 0 0 0 | 0 0 0 0 | 0 0 0 |
| 0 1 0 0 | 1 0 0 0 | 1 0 0 0 | 1 0 0 0 | 1 1 0 0 | 0 0 0 0 | 0 0 0 0 | 0 0 0 0 | 0 0 0 0 | 0 0 0 0 | 0 0 0 |
| 0 1 1 0 | 1 0 0 0 | 0 0 0 0 | 1 0 0 0 | 0 0 0 0 | 1 0 1 0 | 0 0 0 0 | 0 0 0 0 | 0 0 0 0 | 0 0 0 0 | 0 0 0 |
| 0 1 1 1 | 1 0 0 0 | 0 0 0 0 | 0 0 0 0 | 1 0 0 0 | 1 0 0 1 | 0 0 0 0 | 0 0 0 0 | 0 0 0 0 | 0 0 0 0 | 0 0 0 |
| 0 1 0 0 | 1 0 0 0 | 1 0 0 0 | 1 0 0 0 | 1 0 0 0 | 1 0 0 0 | 1 1 0 0 | 0 0 0 0 | 0 0 0 0 | 0 0 0 0 | 0 0 0 |
| 0 0 1 0 | 0 0 0 0 | 1 0 0 0 | 0 0 0 0 | 1 0 0 0 | 0 0 0 0 | 1 0 1 0 | 0 0 0 0 | 0 0 0 0 | 0 0 0 0 | 0 0 0 |
| 0 0 0 1 | 0 0 0 0 | 0 0 0 0 | 1 0 0 0 | 1 0 0 1 | 0 0 0 0 | 0 0 0 0 | 0 0 0 0 | 0 0 0 0 | 0 0 0 0 | 0 0 0 |
| 0 1 0 0 | 1 0 0 0 | 1 0 0 0 | 1 0 0 0 | 1 0 0 0 | 1 1 0 0 | 0 0 0 0 | 0 0 0 0 | 0 0 0 0 | 0 0 0 0 | 0 0 0 |
| 0 1 1 0 | 1 0 0 0 | 0 0 0 0 | 1 0 0 0 | 0 0 0 0 | 1 0 1 0 | 0 0 0 0 | 0 0 0 0 | 0 0 0 0 | 0 0 0 0 | 0 0 0 |
| 0 1 0 1 | 1 0 0 0 | 1 0 0 0 | 0 0 0 0 | 0 0 0 0 | 1 0 0 1 | 0 0 0 0 | 0 0 0 0 | 0 0 0 0 | 0 0 0 0 | 0 0 0 |

**Fig. 2.** 30 linearly independent non-trivial linear combinations of the vectors $K_i$ for the 128-bit variant

As in the 128-bit case, the definition of the vectors $K_0, \ldots, K_{51}$ can be expressed by a system of linear equations:

$$M \cdot \begin{pmatrix} K_0 \\ \vdots \\ K_5 \\ f_6(K_5) \\ \vdots \\ f_{48}(K_{47}) \end{pmatrix} = \begin{pmatrix} K_0 \\ \vdots \\ K_{51} \end{pmatrix} \tag{3}$$

Now, $M$ is a binary matrix of size $52 \times 14$. An exact description of the linear expressions of the vectors $K_i$ can be found in Figure 3 in section 4.

The rank of $M$ is 14 and hence the dimension of the nullspace of $M^t$ is 38. A possible choice of the basis[3] is displayed in Figure 4 in section 4. Again, no

---

[3] I.e., a set of 38 linearly independent non-trivial linear relations of the $K_i$.

| | $K_0$ | | | | | $K_5$ | $f_6(K_5)$ | | | | | | | $\cdots$ $f_{48}(K_{47})$ |
|---|---|---|---|---|---|---|---|---|---|---|---|---|---|---|
| $K_0$ | 1 | 0 | 0 | 0 | 0 | 0 | 0 | 0 | 0 | 0 | 0 | 0 | 0 | 0 |
| | 0 | 1 | 0 | 0 | 0 | 0 | 0 | 0 | 0 | 0 | 0 | 0 | 0 | 0 |
| | 0 | 0 | 1 | 0 | 0 | 0 | 0 | 0 | 0 | 0 | 0 | 0 | 0 | 0 |
| | 0 | 0 | 0 | 1 | 0 | 0 | 0 | 0 | 0 | 0 | 0 | 0 | 0 | 0 |
| $K_4$ | 0 | 0 | 0 | 1 | 0 | 0 | 0 | 0 | 0 | 0 | 0 | 0 | 0 | 0 |
| | 0 | 0 | 0 | 0 | 1 | 0 | 0 | 0 | 0 | 0 | 0 | 0 | 0 | 0 |
| | 1 | 0 | 0 | 0 | 0 | 0 | 1 | 0 | 0 | 0 | 0 | 0 | 0 | 0 |
| | 1 | 1 | 0 | 0 | 0 | 0 | 1 | 0 | 0 | 0 | 0 | 0 | 0 | 0 |
| $K_8$ | 1 | 1 | 1 | 0 | 0 | 0 | 1 | 0 | 0 | 0 | 0 | 0 | 0 | 0 |
| | 1 | 1 | 1 | 1 | 0 | 0 | 1 | 0 | 0 | 0 | 0 | 0 | 0 | 0 |
| | 1 | 1 | 1 | 1 | 1 | 0 | 1 | 0 | 0 | 0 | 0 | 0 | 0 | 0 |
| | 1 | 1 | 1 | 1 | 1 | 1 | 1 | 0 | 0 | 0 | 0 | 0 | 0 | 0 |
| $K_{12}$ | 1 | 0 | 0 | 0 | 0 | 0 | 1 | 1 | 0 | 0 | 0 | 0 | 0 | 0 |
| | 0 | 1 | 0 | 0 | 0 | 0 | 0 | 1 | 0 | 0 | 0 | 0 | 0 | 0 |
| | 1 | 0 | 1 | 0 | 0 | 0 | 1 | 1 | 0 | 0 | 0 | 0 | 0 | 0 |
| | 0 | 1 | 0 | 1 | 0 | 0 | 0 | 1 | 0 | 0 | 0 | 0 | 0 | 0 |
| $K_{16}$ | 1 | 0 | 1 | 0 | 1 | 0 | 1 | 1 | 0 | 0 | 0 | 0 | 0 | 0 |
| | 0 | 1 | 0 | 1 | 0 | 1 | 0 | 1 | 0 | 0 | 0 | 0 | 0 | 0 |
| | 1 | 0 | 0 | 0 | 0 | 0 | 1 | 1 | 1 | 0 | 0 | 0 | 0 | 0 |
| | 1 | 1 | 0 | 0 | 0 | 0 | 1 | 0 | 1 | 0 | 0 | 0 | 0 | 0 |
| $K_{20}$ | 0 | 1 | 1 | 0 | 0 | 0 | 0 | 1 | 1 | 0 | 0 | 0 | 0 | 0 |
| | 0 | 0 | 1 | 1 | 0 | 0 | 0 | 0 | 1 | 0 | 0 | 0 | 0 | 0 |
| | 1 | 0 | 0 | 1 | 1 | 0 | 1 | 1 | 1 | 0 | 0 | 0 | 0 | 0 |
| | 1 | 1 | 0 | 0 | 1 | 1 | 1 | 0 | 1 | 0 | 0 | 0 | 0 | 0 |
| $K_{24}$ | 1 | 0 | 0 | 0 | 0 | 0 | 1 | 1 | 1 | 1 | 0 | 0 | 0 | 0 |
| | 0 | 1 | 0 | 0 | 0 | 0 | 0 | 1 | 0 | 1 | 0 | 0 | 0 | 0 |
| | 0 | 0 | 1 | 0 | 0 | 0 | 0 | 0 | 1 | 1 | 0 | 0 | 0 | 0 |
| | 0 | 0 | 0 | 1 | 0 | 0 | 0 | 0 | 0 | 1 | 0 | 0 | 0 | 0 |
| $K_{28}$ | 1 | 0 | 0 | 0 | 1 | 0 | 1 | 1 | 1 | 1 | 0 | 0 | 0 | 0 |
| | 0 | 1 | 0 | 0 | 0 | 1 | 0 | 1 | 0 | 1 | 0 | 0 | 0 | 0 |
| | 1 | 0 | 0 | 0 | 0 | 0 | 1 | 1 | 1 | 1 | 1 | 0 | 0 | 0 |
| | 1 | 1 | 0 | 0 | 0 | 0 | 1 | 0 | 1 | 0 | 1 | 0 | 0 | 0 |
| $K_{32}$ | 1 | 1 | 1 | 0 | 0 | 0 | 1 | 0 | 0 | 1 | 1 | 0 | 0 | 0 |
| | 1 | 1 | 1 | 1 | 0 | 0 | 1 | 0 | 0 | 0 | 1 | 0 | 0 | 0 |
| | 0 | 1 | 1 | 1 | 1 | 0 | 0 | 1 | 1 | 1 | 1 | 0 | 0 | 0 |
| | 0 | 0 | 1 | 1 | 1 | 1 | 0 | 0 | 1 | 0 | 1 | 0 | 0 | 0 |
| $K_{36}$ | 1 | 0 | 0 | 0 | 0 | 0 | 1 | 1 | 1 | 1 | 1 | 1 | 0 | 0 |
| | 0 | 1 | 0 | 0 | 0 | 0 | 0 | 1 | 0 | 1 | 0 | 1 | 0 | 0 |
| | 1 | 0 | 1 | 0 | 0 | 0 | 1 | 1 | 0 | 0 | 1 | 1 | 0 | 0 |
| | 0 | 1 | 0 | 1 | 0 | 0 | 0 | 1 | 0 | 0 | 1 | 1 | 0 | 0 |
| $K_{40}$ | 0 | 0 | 1 | 0 | 1 | 0 | 0 | 0 | 1 | 1 | 1 | 1 | 0 | 0 |
| | 0 | 0 | 0 | 1 | 0 | 1 | 0 | 0 | 0 | 1 | 0 | 1 | 0 | 0 |
| | 1 | 0 | 0 | 0 | 0 | 0 | 1 | 1 | 1 | 1 | 1 | 1 | 1 | 0 |
| | 1 | 1 | 0 | 0 | 0 | 0 | 1 | 0 | 1 | 0 | 1 | 0 | 1 | 0 |
| $K_{44}$ | 0 | 1 | 1 | 0 | 0 | 0 | 0 | 1 | 1 | 0 | 0 | 1 | 1 | 0 |
| | 0 | 0 | 1 | 1 | 0 | 0 | 0 | 0 | 1 | 0 | 0 | 0 | 1 | 0 |
| | 0 | 0 | 0 | 1 | 1 | 0 | 0 | 0 | 0 | 1 | 1 | 1 | 1 | 0 |
| | 0 | 0 | 0 | 0 | 1 | 1 | 0 | 0 | 0 | 0 | 1 | 0 | 1 | 0 |
| $K_{48}$ | 1 | 0 | 0 | 0 | 0 | 0 | 1 | 1 | 1 | 1 | 1 | 1 | 1 | 1 |
| | 0 | 1 | 0 | 0 | 0 | 0 | 0 | 1 | 0 | 1 | 0 | 1 | 0 | 1 |
| | 0 | 0 | 1 | 0 | 0 | 0 | 0 | 0 | 1 | 1 | 0 | 0 | 1 | 1 |
| | 0 | 0 | 0 | 1 | 0 | 0 | 0 | 0 | 0 | 1 | 0 | 0 | 0 | 1 |

**Fig. 3.** The linear expressions of $K_0, \ldots, K_{51}$ for the 192-bit variant

non-trivial linear relations between the vectors $K_0, K_1, K_2, K_3, K_4, K_5$ and the key $K_{48}, K_{49}, K_{50}, K_{51}$ of the last round exist.

### 3.3    AES with 256 Bit Key Length

Let $\mathcal{K} = (\mathcal{K}_0, \mathcal{K}_1, \mathcal{K}_2, \mathcal{K}_3, \mathcal{K}_4, \mathcal{K}_5, \mathcal{K}_6, \mathcal{K}_7) \in \{0,1\}^{256}$ with $\mathcal{K}_i \in \{0,1\}^{32}$ be the secret key. The description of the key schedule differs from the both given before:

$$K_i := \mathcal{K}_i \qquad\qquad\qquad , i < 8$$
$$K_i := K_{i-8} \oplus f_i(K_{i-1}) , i \ge 8,\ i \bmod 8 = 0$$
$$K_i := K_{i-8} \oplus g(K_{i-1}) \ \ , i \ge 8,\ i \bmod 8 = 4$$
$$K_i := K_{i-8} \oplus K_{i-1} \qquad , i \ge 8,\ i \bmod 8 \notin \{0,4\}$$

Again, $f_i$ and $g$ are non-linear permutations whose exact definitions do not matter.

In fact the description of the key schedule can be simplified. Let $f_i := g$ for $i \bmod 8 = 4$. Then the key schedule can be rewritten to

```
K0        K5  K8                  K16            K24            K32            K40            K48
0 1 1 0 | 0 0 | 1 0 | 0 0 0 0 0 0 0 | 0 1 0|0 0 0 0 | 0 0 0|0 0 1 0 | 0 0 0|0 0 0 0 | 0 1 0|1 0 0 0 | 0 0 0
0 0 0 1 | 0 1 | 0 0 | 0 0 0 0 0 0 0 | 0 1 0|0 0 0 0 | 1 0 0|0 0 1 0 | 0 0 0|1 0 0 0 | 1 0 0|0 0 0 0 | 0 0 0
0 1 0 0 | 0 0 | 1 0 | 0 0 0 0 0 0 0 | 0 1 0|0 0 0 0 | 1 0 0|0 0 1 0 | 0 0 0|1 0 0 0 | 0 1 1|0 0 0 0 | 0 0 0
0 0 1 1 | 0 0 | 0 0 | 0 0 0 1 0 0 0 | 0 1 0|0 0 0 0 | 0 0 0|0 0 0 0 | 0 0 0|1 0 0 0 | 0 1 0|0 1 0 0 | 0 0 0
0 0 0 1 1 0 | 0 0 | 0 0 0 0 0 0 0 | 0 1 0|0 0 0 0 | 0 0 0|0 0 0 0 | 0 0 0|0 0 0 0 | 0 1 0|0 0 1 0 | 0 0 0
0 0 0 1 | 1 | 0 0 | 0 0 0 0 0 0 0 | 0 0 0|0 0 0 0 | 1 0 0|0 0 1 0 | 0 0 0|1 0 0 0 | 0 1 0|0 0 0 1 | 0 0 0
0 1 0 0 | 0 0 | 1 0 | 0 0 0 1 0 0 0 | 0 1 0|0 0 0 0 | 1 0 0|0 0 1 0 | 0 0 0|1 0 0 0 | 0 1 0|0 0 0 0 | 1 1 0 0
0 1 0 0 | 0 0 | 0 0 | 0 0 0 1 0 0 0 | 0 1 0|0 0 0 0 | 1 0 0|0 0 0 0 | 0 0 0|1 0 0 0 | 0 1 0|0 0 0 0 | 1 0 0 1
0 0 1 0 | 0 0 | 1 1 0 | 0 0 0 0 0 0 | 0 0 0|0 0 0 0 | 1 0 0|0 0 0 0 | 0 0 0|0 0 0 0 | 0 1 0|0 0 0 0 | 0 1 0
0 1 1 1 | 1 1 | 1 0 | 0 0 0 1 0 0 0 | 0 0 0|0 0 0 0 | 0 0 0|0 0 0 0 | 0 0 0|0 0 0 0 | 0 0 0|0 0 0 0 | 0 0 0
0 0 1 0 | 0 0 | 0 0 | 0 0 0 1 0 1 0 | 0 0 0|0 0 0 0 | 0 0 0|0 0 0 0 | 0 0 0|0 0 0 0 | 0 0 0|0 0 0 0 | 0 0 0
0 0 1 0 | 1 0 | 0 0 | 0 0 0 1 0 0 0 | 1 0 0|0 0 0 0 | 0 0 0|0 0 0 0 | 0 0 0|0 0 0 0 | 0 0 0|0 0 0 0 | 0 0 0
0 1 1 1 | 0 0 | 1 0 | 1 0 0 0 0 0 0 | 0 0 0|0 0 0 0 | 0 0 0|0 0 0 0 | 0 0 0|0 0 0 0 | 0 0 0|0 0 0 0 | 0 0 0
0 1 1 1 | 1 0 | 1 0 | 0 1 0 0 0 0 0 | 0 0 0|0 0 0 0 | 0 0 0|0 0 0 0 | 0 0 0|0 0 0 0 | 0 0 0|0 0 0 0 | 0 0 0
0 1 1 0 | 0 0 | 1 0 | 1 0 0 0 0 0 0 | 0 0 0|0 0 0 0 | 0 0 0|0 0 0 0 | 0 0 0|0 0 0 0 | 0 0 0|0 0 0 0 | 0 0 0
0 1 0 1 | 0 0 | 1 0 | 0 0 0 1 0 0 1 | 0 0 0|0 0 0 0 | 0 0 0|0 0 0 0 | 0 0 0|0 0 0 0 | 0 0 0|0 0 0 0 | 0 0 0
0 1 0 0 | 0 0 | 1 0 | 0 0 0 1 1 0 0 | 0 0 0|0 0 0 0 | 0 0 0|0 0 0 0 | 0 0 0|0 0 0 0 | 0 0 0|0 0 0 0 | 0 0 0
0 1 0 0 | 0 0 | 1 0 | 0 0 0 1 0 0 0 | 0 1 1|0 0 0 0 | 0 0 0|0 0 0 0 | 0 0 0|0 0 0 0 | 0 0 0|0 0 0 0 | 0 0 0
0 1 1 0 | 0 0 | 1 0 | 0 0 0 0 0 0 0 | 0 1 0|1 0 0 0 | 0 0 0|0 0 0 0 | 0 0 0|0 0 0 0 | 0 0 0|0 0 0 0 | 0 0 0
0 0 1 1 | 0 0 | 1 0 | 0 0 0 0 0 0 0 | 0 1 0|0 1 0 0 | 0 0 0|0 0 0 0 | 0 0 0|0 0 0 0 | 0 0 0|0 0 0 0 | 0 0 0
0 1 0 1 | 0 1 | 1 0 | 0 0 0 1 0 0 0 | 1 0 0|0 0 0 0 | 0 0 0|0 0 0 0 | 0 0 0|0 0 0 0 | 0 0 0|0 0 0 0 | 0 0 0
0 1 0 0 | 1 1 | 1 0 | 0 0 0 1 0 0 0 | 0 1 0|0 0 0 1 | 0 0 0|0 0 0 0 | 0 0 0|0 0 0 0 | 0 0 0|0 0 0 0 | 0 0 0
0 1 0 0 | 0 0 | 1 0 | 0 0 0 1 0 0 0 | 0 1 0|0 0 0 0 | 1 1 0 0|0 0 0 0 | 0 0 0|0 0 0 0 | 0 0 0|0 0 0 0 | 0 0 0
0 0 1 0 | 0 0 | 0 0 | 0 0 0 1 0 0 0 | 0 0 0|0 0 0 1 | 0 1 0|0 0 0 0 | 0 0 0|0 0 0 0 | 0 0 0|0 0 0 0 | 0 0 0
0 0 1 0 | 0 0 | 0 0 | 0 0 0 0 0 0 0 | 0 1 0|0 0 0 0 | 0 0 1|0 0 0 0 | 0 0 0|0 0 0 0 | 0 0 0|0 0 0 0 | 0 0 0
0 0 0 0 | 1 0 | 0 0 | 0 0 0 0 0 0 0 | 0 0 0|0 0 0 0 | 1 0 0|1 0 0 0 | 0 0 0|0 0 0 0 | 0 0 0|0 0 0 0 | 0 0 0
0 1 0 0 | 0 1 | 1 0 | 0 0 0 1 0 0 0 | 0 1 0|0 0 0 0 | 0 0 0|0 1 0 0 | 0 0 0|0 0 0 0 | 0 0 0|0 0 0 0 | 0 0 0
0 0 0 1 | 1 0 | 1 0 | 0 0 0 0 0 0 0 | 0 1 0|0 0 1 0 | 0 0 0|0 0 0 0 | 0 0 0|0 0 0 0 | 0 0 0|0 0 0 0 | 0 0 0
0 1 0 0 | 0 0 | 1 0 | 0 0 0 1 0 0 0 | 0 1 0|0 0 0 0 | 0 0 0|0 0 1 1 | 0 0 0|0 0 0 0 | 0 0 0|0 0 0 0 | 0 0 0
0 1 1 0 | 0 0 | 1 0 | 0 0 0 0 0 0 0 | 0 1 0|0 0 0 0 | 0 0 0|0 0 1 0 | 1 0 0|0 0 0 0 | 0 0 0|0 0 0 0 | 0 0 0
0 1 1 1 | 0 0 | 1 0 | 0 0 0 0 0 0 0 | 0 0 0|0 0 0 0 | 0 0 0|0 0 1 0 | 0 1 0|0 0 0 0 | 0 0 0|0 0 0 0 | 0 0 0
0 1 1 1 | 1 0 | 1 0 | 0 0 0 0 0 0 0 | 0 0 0|0 0 0 0 | 0 0 0|0 0 1 0 | 0 0 1|0 0 0 0 | 0 0 0|0 0 0 0 | 0 0 0
0 0 1 1 | 1 1 | 0 0 | 0 0 0 1 0 0 0 | 0 1 0|0 0 0 0 | 0 0 0|0 0 1 0 | 0 0 1|1 0 0 1 | 0 0 0|0 0 0 0 | 0 0 0
0 1 0 1 | 0 | 1 0 | 0 0 0 1 0 0 0 | 0 0 0|0 0 0 0 | 0 0 0|0 0 1 0 | 0 0 1|1 0 0 0 | 1 0 0|0 0 0 0 | 0 0 0
0 0 1 0 | 1 0 | 0 0 | 0 0 0 1 0 0 0 | 0 0 0|0 0 0 0 | 0 0 0|0 0 1 0 | 0 0 1|1 1 0 0 | 0 0 0|0 0 0 0 | 0 0 0
0 1 0 0 | 0 0 | 1 0 | 0 0 0 1 0 0 0 | 0 1 0|0 0 0 0 | 0 0 0|0 0 1 0 | 0 0 0|1 0 1 0 | 0 0 0|0 0 0 0 | 0 0 0
0 0 1 0 | 0 0 | 0 0 | 0 0 0 1 0 0 0 | 0 0 0|0 0 0 0 | 1 0 0|0 0 1 0 | 0 0 0|0 0 0 0 | 0 0 0
```

**Fig. 4.** 38 linearly independent non-trivial linear combinations of the vectors $K_i$ for the 192-bit variant

$$K_i := K_i \qquad\qquad\qquad, i < 8$$
$$K_i := K_{i-8} \oplus f_i(K_{i-1}) \ , i \geq 8, \ i \bmod 4 = 0$$
$$K_i := K_{i-8} \oplus K_{i-1} \qquad, i \geq 8, \ i \bmod 4 \neq 0$$

This is very similar to the key schedule used in the 128-bit case. Again, each of the key vectors $K_0, \ldots, K_{59}$ can be expressed by a linear combination of the vectors $K_0, \ldots, K_7, f_8(K_7), \ldots, f_{56}(K_{55})$. These can be found in Figure 5 in section 4. The corresponding matrix-vector-product is

$$M \cdot \begin{pmatrix} K_0 \\ \vdots \\ K_7 \\ f_8(K_7) \\ \vdots \\ f_{56}(K_{55}) \end{pmatrix} = \begin{pmatrix} K_0 \\ \vdots \\ K_{59} \end{pmatrix} \tag{4}$$

$M$ has the size $60 \times 21$ and the rank 21. This implies that 39 linearly independent non-trivial linear combinations of the vectors $K_i$ can be found. One possible choice is displayed in Figure 6 in section 4. As in both cases before, the expressions of the vectors $K_0, \ldots, K_7, K_{56}, \ldots, K_{59}$ are linearly independent.

| | $K_0$ | | | | | | | $K_7$ | $f_8(K_7)$ | | | | | | | | | | | | $f_{56}(K_{55})$ |
|---|---|---|---|---|---|---|---|---|---|---|---|---|---|---|---|---|---|---|---|---|---|
| $K_0$ | 1 | 0 | 0 | 0 | 0 | 0 | 0 | 0 | 0 | 0 | 0 | 0 | 0 | 0 | 0 | 0 | 0 | 0 | 0 | 0 | 0 |
| | 0 | 1 | 0 | 0 | 0 | 0 | 0 | 0 | 0 | 0 | 0 | 0 | 0 | 0 | 0 | 0 | 0 | 0 | 0 | 0 | 0 |
| | 0 | 0 | 1 | 0 | 0 | 0 | 0 | 0 | 0 | 0 | 0 | 0 | 0 | 0 | 0 | 0 | 0 | 0 | 0 | 0 | 0 |
| | 0 | 0 | 0 | 1 | 0 | 0 | 0 | 0 | 0 | 0 | 0 | 0 | 0 | 0 | 0 | 0 | 0 | 0 | 0 | 0 | 0 |
| $K_4$ | 0 | 0 | 0 | 0 | 1 | 0 | 0 | 0 | 0 | 0 | 0 | 0 | 0 | 0 | 0 | 0 | 0 | 0 | 0 | 0 | 0 |
| | 0 | 0 | 0 | 0 | 0 | 1 | 0 | 0 | 0 | 0 | 0 | 0 | 0 | 0 | 0 | 0 | 0 | 0 | 0 | 0 | 0 |
| | 0 | 0 | 0 | 0 | 0 | 0 | 1 | 0 | 0 | 0 | 0 | 0 | 0 | 0 | 0 | 0 | 0 | 0 | 0 | 0 | 0 |
| | 0 | 0 | 0 | 0 | 0 | 0 | 0 | 1 | 0 | 0 | 0 | 0 | 0 | 0 | 0 | 0 | 0 | 0 | 0 | 0 | 0 |
| $K_8$ | 1 | 0 | 0 | 0 | 0 | 0 | 0 | 0 | 1 | 0 | 0 | 0 | 0 | 0 | 0 | 0 | 0 | 0 | 0 | 0 | 0 |
| | 1 | 1 | 0 | 0 | 0 | 0 | 0 | 0 | 1 | 0 | 0 | 0 | 0 | 0 | 0 | 0 | 0 | 0 | 0 | 0 | 0 |
| | 1 | 1 | 1 | 0 | 0 | 0 | 0 | 0 | 1 | 0 | 0 | 0 | 0 | 0 | 0 | 0 | 0 | 0 | 0 | 0 | 0 |
| | 1 | 1 | 1 | 1 | 0 | 0 | 0 | 0 | 1 | 0 | 0 | 0 | 0 | 0 | 0 | 0 | 0 | 0 | 0 | 0 | 0 |
| $K_{12}$ | 0 | 0 | 0 | 0 | 1 | 0 | 0 | 0 | 0 | 1 | 0 | 0 | 0 | 0 | 0 | 0 | 0 | 0 | 0 | 0 | 0 |
| | 0 | 0 | 0 | 0 | 1 | 1 | 0 | 0 | 0 | 1 | 0 | 0 | 0 | 0 | 0 | 0 | 0 | 0 | 0 | 0 | 0 |
| | 0 | 0 | 0 | 0 | 1 | 1 | 1 | 0 | 0 | 1 | 0 | 0 | 0 | 0 | 0 | 0 | 0 | 0 | 0 | 0 | 0 |
| | 0 | 0 | 0 | 0 | 1 | 1 | 1 | 1 | 0 | 1 | 0 | 0 | 0 | 0 | 0 | 0 | 0 | 0 | 0 | 0 | 0 |
| $K_{16}$ | 1 | 0 | 0 | 0 | 0 | 0 | 0 | 0 | 1 | 0 | 1 | 0 | 0 | 0 | 0 | 0 | 0 | 0 | 0 | 0 | 0 |
| | 0 | 1 | 0 | 0 | 0 | 0 | 0 | 0 | 0 | 0 | 1 | 0 | 0 | 0 | 0 | 0 | 0 | 0 | 0 | 0 | 0 |
| | 1 | 0 | 1 | 0 | 0 | 0 | 0 | 0 | 1 | 0 | 1 | 0 | 0 | 0 | 0 | 0 | 0 | 0 | 0 | 0 | 0 |
| | 0 | 1 | 0 | 1 | 0 | 0 | 0 | 0 | 0 | 0 | 1 | 0 | 0 | 0 | 0 | 0 | 0 | 0 | 0 | 0 | 0 |
| $K_{20}$ | 0 | 0 | 0 | 0 | 1 | 0 | 0 | 0 | 0 | 1 | 0 | 1 | 0 | 0 | 0 | 0 | 0 | 0 | 0 | 0 | 0 |
| | 0 | 0 | 0 | 0 | 0 | 1 | 0 | 0 | 0 | 0 | 0 | 1 | 0 | 0 | 0 | 0 | 0 | 0 | 0 | 0 | 0 |
| | 0 | 0 | 0 | 0 | 1 | 0 | 1 | 0 | 0 | 1 | 0 | 1 | 0 | 0 | 0 | 0 | 0 | 0 | 0 | 0 | 0 |
| | 0 | 0 | 0 | 0 | 0 | 1 | 0 | 1 | 0 | 0 | 0 | 1 | 0 | 0 | 0 | 0 | 0 | 0 | 0 | 0 | 0 |
| $K_{24}$ | 1 | 0 | 0 | 0 | 0 | 0 | 0 | 0 | 1 | 0 | 1 | 0 | 1 | 0 | 0 | 0 | 0 | 0 | 0 | 0 | 0 |
| | 1 | 1 | 0 | 0 | 0 | 0 | 0 | 0 | 1 | 0 | 0 | 0 | 1 | 0 | 0 | 0 | 0 | 0 | 0 | 0 | 0 |
| | 0 | 1 | 1 | 0 | 0 | 0 | 0 | 0 | 0 | 0 | 1 | 0 | 1 | 0 | 0 | 0 | 0 | 0 | 0 | 0 | 0 |
| | 0 | 0 | 1 | 1 | 0 | 0 | 0 | 0 | 0 | 0 | 0 | 0 | 1 | 0 | 0 | 0 | 0 | 0 | 0 | 0 | 0 |
| $K_{28}$ | 0 | 0 | 0 | 0 | 1 | 0 | 0 | 0 | 0 | 1 | 0 | 1 | 0 | 1 | 0 | 0 | 0 | 0 | 0 | 0 | 0 |
| | 0 | 0 | 0 | 0 | 1 | 1 | 0 | 0 | 0 | 1 | 0 | 0 | 0 | 1 | 0 | 0 | 0 | 0 | 0 | 0 | 0 |
| | 0 | 0 | 0 | 0 | 0 | 1 | 1 | 0 | 0 | 0 | 0 | 1 | 0 | 1 | 0 | 0 | 0 | 0 | 0 | 0 | 0 |
| | 0 | 0 | 0 | 0 | 0 | 0 | 1 | 1 | 0 | 0 | 0 | 0 | 0 | 1 | 0 | 0 | 0 | 0 | 0 | 0 | 0 |
| $K_{32}$ | 1 | 0 | 0 | 0 | 0 | 0 | 0 | 0 | 1 | 0 | 1 | 0 | 1 | 0 | 1 | 0 | 0 | 0 | 0 | 0 | 0 |
| | 0 | 1 | 0 | 0 | 0 | 0 | 0 | 0 | 0 | 0 | 1 | 0 | 0 | 0 | 1 | 0 | 0 | 0 | 0 | 0 | 0 |
| | 0 | 0 | 1 | 0 | 0 | 0 | 0 | 0 | 0 | 0 | 0 | 0 | 1 | 0 | 1 | 0 | 0 | 0 | 0 | 0 | 0 |
| | 0 | 0 | 0 | 1 | 0 | 0 | 0 | 0 | 0 | 0 | 0 | 0 | 0 | 0 | 1 | 0 | 0 | 0 | 0 | 0 | 0 |
| $K_{36}$ | 0 | 0 | 0 | 0 | 1 | 0 | 0 | 0 | 0 | 1 | 0 | 1 | 0 | 1 | 0 | 1 | 0 | 0 | 0 | 0 | 0 |
| | 0 | 0 | 0 | 0 | 0 | 1 | 0 | 0 | 0 | 0 | 0 | 1 | 0 | 0 | 0 | 1 | 0 | 0 | 0 | 0 | 0 |
| | 0 | 0 | 0 | 0 | 0 | 0 | 1 | 0 | 0 | 0 | 0 | 0 | 0 | 1 | 0 | 1 | 0 | 0 | 0 | 0 | 0 |
| | 0 | 0 | 0 | 0 | 0 | 0 | 0 | 1 | 0 | 0 | 0 | 0 | 0 | 0 | 0 | 1 | 0 | 0 | 0 | 0 | 0 |
| $K_{40}$ | 1 | 0 | 0 | 0 | 0 | 0 | 0 | 0 | 1 | 0 | 1 | 0 | 1 | 0 | 1 | 0 | 1 | 0 | 0 | 0 | 0 |
| | 1 | 1 | 0 | 0 | 0 | 0 | 0 | 0 | 1 | 0 | 0 | 0 | 1 | 0 | 0 | 0 | 1 | 0 | 0 | 0 | 0 |
| | 1 | 1 | 1 | 0 | 0 | 0 | 0 | 0 | 1 | 0 | 0 | 0 | 0 | 0 | 1 | 0 | 1 | 0 | 0 | 0 | 0 |
| | 1 | 1 | 1 | 1 | 0 | 0 | 0 | 0 | 1 | 0 | 0 | 0 | 0 | 0 | 0 | 0 | 1 | 0 | 0 | 0 | 0 |
| $K_{44}$ | 0 | 0 | 0 | 0 | 1 | 0 | 0 | 0 | 0 | 1 | 0 | 1 | 0 | 1 | 0 | 1 | 0 | 1 | 0 | 0 | 0 |
| | 0 | 0 | 0 | 0 | 1 | 1 | 0 | 0 | 0 | 1 | 0 | 0 | 0 | 1 | 0 | 0 | 0 | 1 | 0 | 0 | 0 |
| | 0 | 0 | 0 | 0 | 1 | 1 | 1 | 0 | 0 | 1 | 0 | 0 | 0 | 0 | 0 | 1 | 0 | 1 | 0 | 0 | 0 |
| | 0 | 0 | 0 | 0 | 1 | 1 | 1 | 1 | 0 | 1 | 0 | 0 | 0 | 0 | 0 | 0 | 0 | 1 | 0 | 0 | 0 |
| $K_{48}$ | 1 | 0 | 0 | 0 | 0 | 0 | 0 | 0 | 1 | 0 | 1 | 0 | 1 | 0 | 1 | 0 | 1 | 0 | 1 | 0 | 0 |
| | 0 | 1 | 0 | 0 | 0 | 0 | 0 | 0 | 0 | 0 | 1 | 0 | 0 | 0 | 1 | 0 | 0 | 0 | 1 | 0 | 0 |
| | 1 | 0 | 1 | 0 | 0 | 0 | 0 | 0 | 1 | 0 | 1 | 0 | 0 | 0 | 0 | 0 | 1 | 0 | 1 | 0 | 0 |
| | 0 | 1 | 0 | 1 | 0 | 0 | 0 | 0 | 0 | 0 | 1 | 0 | 0 | 0 | 0 | 0 | 0 | 0 | 1 | 0 | 0 |
| $K_{52}$ | 0 | 0 | 0 | 0 | 1 | 0 | 0 | 0 | 0 | 1 | 0 | 1 | 0 | 1 | 0 | 1 | 0 | 1 | 0 | 1 | 0 |
| | 0 | 0 | 0 | 0 | 0 | 1 | 0 | 0 | 0 | 0 | 0 | 1 | 0 | 0 | 0 | 1 | 0 | 0 | 0 | 1 | 0 |
| | 0 | 0 | 0 | 0 | 1 | 0 | 1 | 0 | 0 | 1 | 0 | 1 | 0 | 0 | 0 | 0 | 0 | 1 | 0 | 1 | 0 |
| | 0 | 0 | 0 | 0 | 0 | 1 | 0 | 1 | 0 | 0 | 0 | 1 | 0 | 0 | 0 | 0 | 0 | 0 | 0 | 1 | 0 |
| $K_{56}$ | 1 | 0 | 0 | 0 | 0 | 0 | 0 | 0 | 1 | 0 | 1 | 0 | 1 | 0 | 1 | 0 | 1 | 0 | 1 | 0 | 1 |
| | 1 | 1 | 0 | 0 | 0 | 0 | 0 | 0 | 1 | 0 | 0 | 0 | 1 | 0 | 0 | 0 | 1 | 0 | 0 | 0 | 1 |
| | 0 | 1 | 1 | 0 | 0 | 0 | 0 | 0 | 0 | 0 | 1 | 0 | 1 | 0 | 0 | 0 | 0 | 0 | 1 | 0 | 1 |
| | 0 | 0 | 1 | 1 | 0 | 0 | 0 | 0 | 0 | 0 | 0 | 0 | 1 | 0 | 0 | 0 | 0 | 0 | 0 | 0 | 1 |

**Fig. 5.** The linear expressions of $K_0, \ldots, K_{59}$ for the 256-bit variant

## 4    Exact Descriptions

In this section, we provide an exact description of

- the expression of $K_i$ as a linear combination of the vectors $K_0, \ldots, K_{N_k-1}$ and $f_j(K_{j-1})$
- a basis for all non-trivial linear combinations of the vectors $K_0, \ldots, K_{4N_r+3}$

for all three AES variants.

We demonstrate on an example how the tables have to be read. Figure 1 shows how $K_0, \ldots, K_{43}$ (in the 128-bit case) can be expressed by a linear combination of $K_0, K_1, K_2, K_3, f_4(K_3), \ldots, f_{40}(K_{39})$. Assume for example that we are interested

| $K_0$ | | | | | $K_7$ | $K_8$ | | | | | | | | $K_{20}$ | | | | | | | | $K_{32}$ | | | | | | | | $K_{44}$ | | | | | | | | $K_{56}$ | | | |
|---|---|---|---|---|---|---|---|---|---|---|---|---|---|---|---|---|---|---|---|---|---|---|---|---|---|---|---|---|---|---|---|---|---|---|---|---|---|---|---|---|---|

Fig. 6. 39 linearly independent non-trivial linear combinations of the vectors $K_i$ for the 256-bit variant

in the expression of $K_{19}$. This can be found in the 20th row of the table (not counting the "indexing row" at the top). This row contains only two entries unequal zero: one in the 4th column which corresponds to $K_3$ and one in the 8th column which corresponds to $f_{16}(K_{15})$. This means that

$$K_{19} = K_3 \oplus f_{16}(K_{15}).$$

Figure 2 displays 30 linearly independent non-trivial linear relations of $K_0, \ldots, K_{43}$. For example, the last but four row shows that

$$K_1 \oplus K_2 \oplus K_4 \oplus K_{12} \oplus K_{20} \oplus K_{22} = 0.$$

## 5    Conclusion

The current paper gives a complete description of *all* linear relationships between singular round key values $K_i \in \{0,1\}^{32}$. More specifically, for each key schedule a matrix $M$ is described such that each linear relationship of the form

$$\bigoplus_{i=0}^{4N_r+3} c_i \cdot K_i = 0, \quad c_i \in \mathbb{F}_2$$

corresponds to a vector in the nullspace of $M^t$. Such relationships can in principle be useful for related-key attacks against the AES.

Our observations are independent from the choice of the nonlinear function $f$ (in fact, we could even allow independent nonlinear functions for each of the $f_i$).

If the AES key schedule would evaluate one nonlinear function $f_i$ *for each* round key value $K_i$ ($i \geq 4$ for 128-bit keys ect.), the corresponding matrix $M$ would be a *square matrix* of *full rank*, and thus no useful linear relationships could exist.

This would, however, decrease the performance of the AES key schedule significantly, without solving an immediate problem: We could verify that no *exclusive* relationship between the round key values from the first round and the last round exists. Thus, there is no straightforward way to exploit of our findings to mount a related key attack against the AES.

## Acknowledgment

The author would like to thank Joe Cho, Nicolas Courtois, Erik Zenner, Matthias Krause and the unknown referees for helpful comments and discussions.

## References

1. Nicolas Courtois, Private Communication.
2. J. Daemen and V. Rijmen: *The Design of Rijndael*, 2002, Springer.
3. Niels Ferguson, John Kelsey, Stefan Lucks, Bruce Schneier, Mike Stay, David Wagner, and Doug Whiting. Improved Crypanalysis of Rijndael. *Fast Software Encryption 2000*, Springer Lecture Notes in Computer Science.

# The Inverse S-Box, Non-linear Polynomial Relations and Cryptanalysis of Block Ciphers[*]

Nicolas T. Courtois

Axalto Cryptographic Research & Advanced Security,
36-38 rue de la Princesse, BP 45, 78430 Louveciennes Cedex, France
courtois@minrank.org
http://www.nicolascourtois.net

**Abstract.** This paper is motivated by the design of AES. We consider a broader question of cryptanalysis of block ciphers having very good non-linearity and diffusion. Can we expect anyway, to attacks such ciphers, clearly designed to render hopeless the main classical attacks ? Recently a lot of attention have been drawn to the existence of multivariate algebraic relations for AES (and other) S-boxes. Then, if the XSL-type algebraic attacks on block ciphers [10] are shown to work well, the answer would be positive. In this paper we show that the answer is certainly positive for many other constructions of ciphers. This is not due to an algebraic attack, but to new types of generalised linear cryptanalysis, highly-nonlinear in flavour. We present several constructions of somewhat special practical block ciphers, seemingly satisfying all the design criteria of AES and using similar S-boxes, and yet being extremely weak. They can be generalised, and evolve into general attacks that can be applied - potentially - to any block cipher.

**Keywords:** Block ciphers, AES, Rijndael, interpolation attack on block ciphers, fractional transformations, homographic functions, multivariate equations, Feistel ciphers, generalised linear cryptanalysis, bi-linear cryptanalysis.

## 1   Introduction

AES (Rijndael) [14, 15] is a rather accomplished realisation of certain philosophy that culminates two decades of research in the design of modern block ciphers. It has important security margins and at present attacking full Rijndael is very ambitious. The research on AES focuses rather on better understanding its security by following two paths. First approach analyses the security of reduced-round versions of AES against known attacks. Another line of research is to attack, instead of Rijndael, its design principles. Though the outcome of this approach will in most cases give results not being directly applicable to AES, it remains

---

[*] Work supported by the French Ministry of Research RNRT Project "X-CRYPT".

H. Dobbertin, V. Rijmen, A. Sowa (Eds.): AES 2004, LNCS 3373, pp. 170–188, 2005.

extremely interesting. This is because Rijndael pushes some of these design principles such as high non-linearity or good diffusion to their theoretical limits, thus giving us the opportunity and motivation to explore these limits, and uncover possible pitfalls (are they serious or not).

For example the resistance of the Inverse function in $GF(2^n)$ to linear, differential and higher-order differential attacks is exceptional and close to optimality, see [4]. On page 6 of [16], the designers of AES say: *"[...] The disadvantage of these boxes is that they have a simple description in $GF(2^m)$, which is also the field in which the diffusion layer is linear. This may create uneasy feelings, but we are not aware of any vulnerability caused by this property. For the time being we challenge cryptanalysts to demonstrate any vulnerability caused by this property. Should such a vulnerability exist, one can always replace the Sboxes by Sboxes with similar properties, that are not algebraic over $GF(2^m)$. [...]"*

Unfortunately an important vulnerability of the inverse S-box does exist. It follows the line of research that has already been around for some time now. Historically the idea goes back to cryptanalysis of some rather esoteric public key schemes by Patarin [29], greatly improved by Courtois, Meier et al. [7, 13], and followed without proper acknowledgment by Faugère and Joux [24]. The seminal idea (due to Patarin) is to study the security of a cipher component not in terms of Boolean/algebraic functions, but in terms of Boolean/algebraic **relations** that involve both inputs and output bits. This idea is very powerful, and in the last two years, it has led to a sudden collapse of several important families of stream ciphers, as demonstrated by Courtois et al in [11, 12, 1, 8, 9] and numerous other recent papers. But does it matter at all for block ciphers ?

An early warning has been issued by Jakobsen at Crypto'98 [22]. He proposes attacks on block ciphers based on univariate (and tentatively also multivariate) polynomial approximations, and already speaks explicitly of using (probabilistic) algebraic relations. Jakobsen clearly makes his point showing that to obtain secure ciphers *"[...] it is not enough that round functions have high Boolean complexity. Likewise, good properties against differential and linear attacks are no guarantee either. In fact, many almost prefect non-linear functions should be avoided exactly because they are too simple algebraically [...]"*. Yet Jakobsen did not propose neither really surprising nor really devastating attacks, and so far his results are rather seen as a very special case of Generalised Linear Cryptanalysis (GLC) that breaks badly some very special ciphers and has no implication whatsoever for all the other ciphers.

Each Rijndael S-box, though very complex when regarded as a function, can be characterised in several ways by algebraic relations, cf. [10, 27], being true with very high probability, usually 1. When the XL attack was first introduced by Courtois, Klimov, Patarin and Shamir [30], it became sensible to combine these ideas and to write the problem of recovering the AES key as solving a system of multivariate quadratic equations. This seems, at first sight, rather extensively stupid, as obviously we are facing an NP-complete problem, and any other cipher can be attacked in a similar way. Even though for AES the system is somewhat over-determined, and even with an optimistic evaluation of XL, there is clearly no hope to get an attack faster than $2^{300}$. Yet with time, this idea appeared less and less stupid. Courtois and Pieprzyk proposed a method called XSL [10] that allows to substantially lower the (still naive) complexity estimation of an algebraic attack, by adapting the basic idea of XL to the sparsity and the specific structure of these equations. Then Murphy and Robshaw followed [27] with an (in theory) equivalent version of the same approach, writing quadratic equations over $GF(256)$ instead of $GF(2)$, yet yielding more sparsity, and giving hopes for even faster attacks. Thus a rather outrageous idea appeared: this version of XSL attack appears (in first, very naive estimations) to have a potential to recover an AES key in less than $2^{128}$ AES computations, given only one single known plaintext. So far the real feasibility of such attacks is far from being clear.

This paper is also exploiting, very loosely speaking the same, vulnerabilities of the inverse function in $GF(2^n)$ function. As Jakobsen in [22] we will work on multivariate/univariate approximations and relations. Our goal is not to propose attacks on AES, it is more educational: we wish to demonstrate that there are ciphers that have very high-nonlinearity and exceptionally good diffusion but can be broken in practice even for a very large number of rounds. We will make extensive use of the inverse function in $GF(2^n)$ and some of its linear equivalents, both as a component for building highly insecure ciphers, and (quite surprisingly) as the algebraic structure that can be exploited in attacks.

In this paper we follow the following methodology: first we develop some results about composing functions with various operations. Then construct weak contrived ciphers incorporating the inverse in $GF(2^n)$, thus looking secure and satisfying all the known design criteria. Finally from here we develop general families of attacks applicable to (more or less) any cipher. The paper is composed of two rather independent parts. In the first part we study, improve and propose attacks that are based on univariate algebraic equations/relations. These are applied to propose weak Substitution-Permutation Network ciphers (SPN), and will lead to describing a new very general class of attacks on such ciphers. In the second part we are rather concerned with Feistel ciphers and we will study attacks based on bi-variate and multivariate equations.

## Part I - Insecure Substitution-Permutation Networks

## 2    Whitening Ciphers and Known Weak Constructions

We define a subclass of Substitution-Permutation Network (SPN) ciphers that we call Whitening Ciphers. There are ciphers in which we alternatively XOR the state with some derived key $K_i \in GF(2^n)$, and some function $F : GF(2^n) \rightarrow GF(2^n)$ that does not depend on the key and in most cases (but not always) will be bijective and identical for every round.

Serpent [3] and Rijndael [14] are perfect examples of whitening ciphers. Many other ciphers, for example DES and other Feistel ciphers that use XOR to combine consecutive keys, can still be seen as whitening ciphers. In this case $F$ is not bijective, a different $F$ is used in the first and the last round, the intermediate data are redundant, and the key schedule is weak - some bits are always 0.

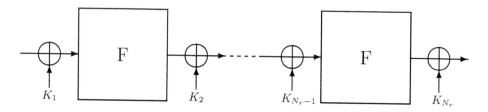

**Fig. 1.** Whitening Ciphers with identical round function $F$

We start with some simple examples of insecure whitening ciphers.

## 2.1     When $F$ Is of Low Degree

For example $n = 128$, $F$ is a polynomial of degree 3, and the number of rounds is $N_r = 16$. Then the whole encryption function is a fixed key-dependent polynomial of degree $D = 3^{16} \approx 2^{25}$. This is not a big degree.

The resulting attack is called an *instance deduction* (cf. [18]) - it does not recover the key but a partial equivalent of it, for example a formula that allows to decrypt/encrypt a certain fraction of messages. Naively the full polynomial can be recovered by Gaussian reduction given $D$ known plaintexts. Alternatively, Jakobsen and Knudsen show on page 5 in [20]. that it can be done much faster in time essentially $D$, given $D$ chosen plaintexts, In [23] it was claimed that it should be done in time $D$ even with $D$ *known* plaintexts, but no proof is given and the result does not seem to be true. At any rate, when $D = 3^{16} \approx 2^{25}$, at least the chosen plaintext attack can be handled in less than 1 second on a PC.

**Remark:** In practice, if we want to build such a cipher, an additional problem will be to have $F$ that is bijective. When $n$ is odd, $GCD(3, 2^n - 1) = 1$ and $X \mapsto X^3$ can be used. Other solutions are known: Dickson polynomials. For example when $2^n \equiv 2 \bmod 5$ or $2^n \equiv 3 \bmod 5$, the polynomial $X^5 + X^3 + 1$ is a permutation, see Section 7 of [25].

## 2.2     When $F$ Is Approximated by a Function of Low Degree

Then, following Jakobsen, the cipher is still insecure. When the whole cipher can be approximated with a polynomial of degree $D^{N_r}$ true with probability $\varepsilon^{N_r}$, then following [22], we may apply Sudan's algorithm to recover this equation, with a complexity of these attacks being a low degree polynomial in $\frac{D^{N_r}}{\varepsilon^{2N_r}}$.

**Remark:** These attacks by polynomial interpolation can be declined in two versions. It can be exploited as a known plaintext attack, and in this case it is a special case of Generalised Linear Cryptanalysis (GLC). It can also be exploited in a chosen plaintext attack, and in this case it becomes a special case of Higher-order Differential Attacks, see [20]. The second variant is less frequently applicable in practice, but the complexity to recover the polynomial should be lower (at least in the non-noisy case, as discussed above). Moreover for many practical ciphers, there is no need at all for recovering the polynomial even in the noisy case (!). We can use a differential of some order that makes the polynomial vanish, to detect the noisy polynomial approximation without recovering it, and build a distinguisher on $N_r - 1$ rounds that should allow key recovery for the full cipher by guessing some relevant key bits in the last round.

**What's Next.** So far we only have serious, but marginal attacks that operate only on contrived (weak) ciphers. Later we will propose more general constructions of insecure ciphers, that can simultaneously incorporate several different components. At some point we will we obtain a completely general attack that applies (potentially) to any cipher, contains the Jakobsen attacks described above, contains the linear cryptanalysis, and - importantly - does break the barrier of linear/low degree approximations.

## 3    Insecure Ciphers Based on the Rijndael S-Box

Some functions became favorite components for designing block ciphers because they are anything but linear/low degree approximations. Let $F = Inv$ be the full-size inverse function with the usual $0 \mapsto 0$:

$$Inv(X) = \begin{cases} X^{-1} & \text{in } GF(2^n) \text{ if } X \neq 0 \\ 0 & \text{otherwise} \end{cases}$$

For now, it is full size, and the representation of the finite field is assumed to be known (later we will consider hidden fields of unknown size).

We call an Almost-Invariant, any property that is invariant but remains true with a slowly decreasing probability. We have the following theorem:

**Theorem 3.0.1 (Almost-Invariant Polynomial Relation Attack).**
For any cipher $X \mapsto Y = E_K(X)$ that composes in any order

(a) $N_r$ applications of $Inv$ in $GF(2^n)$,
(b) any number of XORs with different subkeys or constants,
(c) any number of multiplications by a subkey or a constant, must be $\neq 0$,

there exist $(\alpha, \beta, \gamma, \delta) \in GF(2^n)^4$ such that:

$$\mathbb{P}_{X \in GF(2^n)} [\alpha XY + \beta X + \gamma Y + \delta = 0 \mid Y = E_K(X)] \geq \left(1 - \frac{1}{2^n}\right)^{N_r} \geq \left(1 - \frac{N_r}{2^n}\right)$$

**Proof.** The proof is done by induction. We assume that for a cipher with $N_r$ rounds, there is an equation of the form

$$\alpha XY + \beta X + \gamma Y + \delta = 0$$

that holds for the fraction $\left(1 - \frac{1}{2^n}\right)^{N_r}$ of all $(X, Y) = (X, E_K(X))$. The cases of adding one addition or one multiplication are obvious, the equation still exists.

What if we add an *Inv* ? Let $T = \begin{cases} Y^{-1} & \text{if } Y \neq 0 \\ 0 & \text{otherwise} \end{cases}$.

We multiply the equation above by $T$:

$$\alpha TXY + \beta TX + \gamma TY + \delta T = 0$$

We have $TY = 1$ for all $X$ except when $Y = 0$. The resulting equation will be true with probability that gets multiplied by $\left(1 - \frac{1}{2^n}\right)$.

$$\beta TX + \alpha X + \delta T + \gamma = 0$$

The probability decreases. (This equation will probably not be true when $X$ is such that $Y = 0$, and $T = 0$, except if $\alpha X + \gamma = 0$ which is in general not true.) This ends the proof. $\qquad\square$

**Remark:** When $N_r$ is small, a special case of this theorem exists in another, different form, see equation (6) on page 11 of Jakobsen and Knudsen paper [20]. Yet as it stands, the result of [20] is false and below we give a generalised and corrected version of it. It can also be seen as another, equivalent (but seemingly better) formulation of Theorem 3.0.1:

**Theorem 3.0.2 (Homographic Approximation Version of Thm. 3.0.1).** For any cipher $X \mapsto Y = E_K(X)$ that composes in any order

(a) $N_r$ applications of *Inv* in $GF(2^n)$,
(b) any number of XORs with different subkeys or constants,
(c) any number of multiplications by a subkey or a constant, must be $\neq 0$,

there exist $(\alpha, \beta, \gamma, \delta) \in GF(2^n)^4$ such that:

$$\mathbb{P}_{X \in GF(2^n)} \left[ Y = \frac{\alpha X + \beta}{\gamma X + \delta} \mid Y = E_K(X) \right] \geq \left(1 - \frac{1}{2^n}\right)^{N_r} \geq \left(1 - \frac{N_r}{2^n}\right)$$

**Proof.** The proof is obvious and even easier than for Theorem 3.0.1 above.

**Comparison to [20].** First of all, our result is more general compared to (6) in [20] that does not include the "multiplying by a constant" case (c). Moreover, there are two flaws in the result of [20]: the authors assume that $\alpha \neq 0$ and they neglect the singularity problem of *Inv* claiming that the approximation would be true all the time. Both these omissions are not serious when the number of round is small $N_r \ll 2^n$ and in this case the result (6) of [20] is true with good probability - for ciphers combining only elements of type (a) and (b). Otherwise, the correct result is our Theorem 3.0.2 or Theorem 3.0.1.

# 4    Composition and Approximation Properties of *Inv*

As we will see now, the theory of Rijndael *Inv* ii rich and non-trivial.

## 4.1    Homographic Functions

In mathematics the functions of the form $X \mapsto \frac{\alpha X + \beta}{\gamma X + \delta}$ are called *homographic functions* or *linear fractional transformations*. It is well known that they can be represented by $2 \times 2$ matrices $\begin{pmatrix} \alpha & \beta \\ \gamma & \delta \end{pmatrix}$. The composition of these functions is equivalent to multiplying their matrices (!).

A cross-ratio of 4 pairwise different points $R(t, u, v, w) = \frac{t-u}{t-w} / \frac{v-u}{v-w}$ is known to be an invariant for such transformations The cross-ratio can therefore be used in cryptanalysis as suggested by Vaudenay and Aoki, see Section 2.4. of [2]. However, for cryptanalysis of compositions such as in Theorems 3.0.2 and 3.0.1, with *Inv* version of the inverse, it is again an almost-invariant, not an invariant.

## 4.2    What Is the Difference Between *Inv* and the Inverse ?

More precisely, it will be invariant as long as we do not encounter any singularity in which 0 is mapped to 0. Thus we have:

1. $R(a+k, b+k, c+k, d+k) = R(a, b, c, d)$,
2. $R(\mu a, \mu b, \mu c, \mu d) = R(a, b, c, d)$ for $\mu \neq 0$,
3. $R(1/a, 1/b, 1/c, 1/d) = R(a, b, c, d)$ for non-zero elements,
4. $R(1/a, 1/b, 1/c, 0) = a/c \cdot R(a, b, c, 0)$.

This causes a discontinuity in the invariant.

We see that, the function *Inv* of Rijndael is **not** strictly speaking a homographic function. It is equal to functions of the form $X \mapsto \frac{\alpha X + \beta}{\gamma X + \delta}$ except in one point, when 0 is mapped to 0. It is possible to see that this "completion" is the reason why in Theorems 3.0.2 and 3.0.1. the probability **does** decrease with the number of rounds. Functions that we get by composition are less and less homographic and this is why we talk about *homographic approximation*.

This "completion" with $0 \mapsto 0$ has important and non-trivial properties. There are three ways of defining the inverse function for a finite field:

1. We can have a bijection on 255 elements $GF(256)^* \to GF(256)^*$, thus avoiding the pole. This is acceptable as long as inversion alone is concerned, but in general makes it impossible composing with other homographic functions to form a group: they would have poles at different places, and functions with a different domain can not easily be composed.
2. We can have a bijection on 257 elements $\overline{GF(256)} \to \overline{GF(256)}$ with $\overline{GF(256)} = GF(256) \cup \{\infty\}$. For example the inverse will be defined as:

$$\overline{Inv}(X) = \begin{cases} X^{-1} & \text{if } X \notin \{0, \infty\} \\ 0 \mapsto \infty \\ \infty \mapsto 0 \end{cases}.$$

This is an eminently interesting version. We can compose this function with other homographic functions defined as follows:

$$
\overline{Inv}(X) = \begin{cases} \frac{\alpha X + \beta}{\gamma X + \delta} & \text{if } X \notin \{-\frac{\delta}{\gamma}, \infty\} \\ -\frac{\delta}{\gamma} \mapsto \infty \\ \infty \mapsto \frac{\alpha}{\gamma} \end{cases}.
$$

and we still get a homographic transformation. We obtain a group.

3. We can have a bijection on 256 elements $Inv : GF(256) \rightarrow GF(256)$ that is used in Rijndael. It is important to note that $Inv$ can be seen as as restriction to $GF(256)$ of $\overline{Inv} \circ \tau$, with $\tau$ being a functions that swaps $\infty$ and 0, leaving all the other points unchanged. This "swap" that occurs in $Inv$ when we put $0 \mapsto 0$ explains why when we compose $Inv$ with key additions, the composition function will be homographic with decreasing probability as shown by Theorems 3.0.2 and 3.0.1.

## 4.3    Big and Small Groups, Approximation, AES and DES

The homographic function $1/X$ over $\overline{GF(256)}$ composes well with constant/key additions and with constant/key multiplications to form a quite small group. What is the group generated when we replace $1/X$ par $Inv$ ? With good probability, we still get homographic functions, but the approximation probability decreases with the number of rounds. The answer is non-trivial and rather surprising. The group generated is the group of all permutations. Moreover, the result remains true also when we do not use multiplications and compose only $Inv$ with additions of constants. We have the following result:

**Theorem 4.3.1 (The Group Generated by $Inv$ and XORs).**
The group generated by composing $Inv$ and constant/key additions is exactly the group of all permutations of $GF(2^n)$.

The proof is given in Appendix A.

## Consequences for Block Cipher Cryptanalysis

This fact is very closely related to the famous question whether DES is a group or not. If DES were a group it would have serious consequences on the security of DES, because it means that triple DES would not really be more secure than DES. Luckily it was concluded that DES is not a group [5]. However, as we will explain now, the question of such attacks on triple DES remains widely open. Assume that, as for Theorem 4.3.1 compared to Theorem 3.0.2, DES is not a group, but yet each 64-bit permutation obtained with DES can be seen as equal with some probability, to an element of some group $G$. Then even if this probability were quite small, (e.g. $2^{-10}$), we would obtain an attack on triple DES as follows. We try to guess a presumably 56-bit (maybe less) information that characterise the group element $g \in G$ expected to approximate our triple

DES instance. Given a pair $(P, C)$ produced with the triple-DES, the property $C = g(P)$ will be satisfied with probability $2^{-30}$. A random permutation will satisfy it with a negligible probability of about $2^{-64}$. Thus the right $g$ can be detected in practice. We get a practical distinguishing attack requiring about $2^{56}$ computations and $\mathcal{O}(2^{30})$ known plaintexts.

We conclude that it is not enough to show that DES or AES is not a group. One should design block ciphers in such a way that the encryption operations (or single round operations) should not belong to a small group, and should not have "good" approximations by elements of some group that is not too large.

## 5     More Constructions of Insecure SPN Ciphers

With Theorem 3.0.1, we (already) get a very interesting result. We compose $Inv$'s, multiplication by some (key-dependent or not) constants and XORs by other constants. We get a family of block ciphers such that:

1. It is stable by composition of a few ciphers (but not for a big number).
2. Since they apply the inverse to the whole state, they are strongly resistant to linear, differential and higher-order differential attacks, see [4].
3. The security of these ciphers (and also for more general ones we will propose later) does **not** grow exponentially with the number of rounds:
   (a) When the number of rounds is exponential, up to about $2^n$, for example $N_r = 2^n$, these ciphers can be easily broken given $\mathcal{O}(1)$ known plaintexts. If our goal is only to distinguish the cipher from random, we can use the cross-ratio (cf. Section 4.2 or [2]). For any 4-tuple of known plaintexts it should be invariant with good probability. Otherwise, we use the equation of Theorem 3.0.1 that is true with large probability being at least $(1 - 1/2^n)^{2^n} \approx 1/e$. Any subset of four known plaintext allows to recover the equation with good probability, that is then checked: should remain valid for an important fraction of other plaintexts. We get an instance deduction attack that recovers the $(\alpha, \beta, \gamma, \delta) \in GF(2^n)^4$ and uses them to encrypt/decrypt any message with good probability.
   (b) However, when the number of rounds is very large, some of these ciphers are **provably secure without any assumption**. This is an immediate corollary of Theorem 4.3.1.

### 5.1     Combining with Polynomial Equations

We will combine our class of insecure ciphers with the Jakobsen attack of [22]: we allow also components that are polynomials of small degree.

**Theorem 5.1.1 (Higher-Degree Homographic Approximation Attack).** For any cipher $X \mapsto Y = E_K(X)$ that mixes:

(a) $N_r$ applications of $Inv$ in $GF(2^n)$,
(b) any number of XORs with different subkeys or constants,

(c) any number of multiplications by a subkey or a constant, must be $\neq 0$,

(d) small number of rounds that are small degree polynomials with the total product of their degrees being $D$.

(e) all these combined with noise, or equivalently we assume that all the components of the cipher are not (a-d) but equal to such with some probability $\varepsilon_i$, and with total combined approximation probability being $\varepsilon = \prod \varepsilon_i$.

Then there exist two polynomials $P(X)$ and $Q(X)$ of degree $D$ such that:

$$\mathbb{P}_{X \in GF(2^n)} \left[ Y = \frac{P(X)}{Q(X)} \mid Y = E_K(X) \right] \geq \varepsilon \left( 1 - \frac{1}{2^n} \right)^{N_r} \geq \varepsilon \left( 1 - \frac{N_r}{2^n} \right)$$

**Resulting Attack:** The existence of such polynomial relations can be efficiently checked with a bivariate version of the Sudan's Algorithm, see [22].

**Proof of Theorem 5.1.1.** Again the proof is done by induction and we need to verify step by step that all transformations preserve the property, with degree of the polynomials increasing multiplicatively in the case (d). The cases (a) and (c) are completely trivial. For (b) we write: $P(X)/Q(X) + C = (P(X) + C \cdot Q(X))/Q(X)$. For (d) we observe that if we apply a polynomial of degree $D_1$ $A = a_0 \ldots a_{D_1} X^{D_1}$ to a fraction $(P(X)/Q(X))$ of two polynomials of degree $D_2$ the result can be written as:

$$\frac{a_0 Q(X)^{D_1} + a_1 P(X)^1 Q(X)^{D_1-1} + \ldots a_{D_1} P(X)^{D_1}}{Q(X)^{D_1}}.$$

Clearly a fraction of two polynomials of degree $D_1 D_2$. This ends the proof. $\square$

## 5.2 Further Extension

Our class of insecure ciphers (and our attack) can be extended by using automorphisms of the finite field $GF(2^n)$. This allows to encompass more linear equivalents of the inverse function (so far we used only $X \mapsto a/X$). It also allows to include polynomials that are of high degree without increasing the final $D$.

**Theorem 5.2.1 (Extended Higher-Degree Homographic Attack).**
For any cipher $X \mapsto Y = E_K(X)$ that mixes:

(a) $N_r$ applications of $Inv$ in $GF(2^n)$,

(b) any number of XORs with different subkeys or constants,

(c) any number of multiplications by a subkey or a constant, must be $\neq 0$,

(d) small number of rounds that are small degree polynomials with the total product of their degrees being $D$,

(e) any number of squares in $GF(2^n)$,

(f) all these combined with noise, or equivalently we assume that all the components of the cipher are not (a-d) but equal to such with some probability $\varepsilon_i$, and with total combined approximation probability being $\varepsilon = \prod \varepsilon_i$.

Then there exist $r \in \mathbb{N}$ and two polynomials $P(X)$ and $Q(X)$ of degree $D$ such that:

$$\mathbb{P}_{X \in GF(2^n)} \left[ Y = \frac{P(X^{2^r})}{Q(X^{2^r})} \mid Y = E_K(X) \right] \geq \varepsilon \left( 1 - \frac{1}{2^n} \right)^{N_r} \geq \varepsilon \left( 1 - \frac{N_r}{2^n} \right)$$

**Proof.** The proof is nearly the same. We observe that the Frobenius automorphism (e) commutes with all the other operations (a-d), except it replaces constants/subkeys by a different constant, and polynomials by a different polynomial. Therefore we can safely put all squares at the beginning and we get the result from Theorem 5.1.1.    □

**Resulting Attack:** It is still possible to check if such equation exists, we guess $r \in \{0, n-1\}$ and proceed with a version of Sudan's Algorithm cf. [22].

# 6    New General Attack on Whitening Ciphers

Now we will introduce a new very general attack that can be applied to potentially any whitening cipher (and even to other ciphers). The idea is as follows: consider 128-bit whitening cipher. It is very unlikely that it has any kind of equational property such as in Theorem 5.2.1. However if we select some, say 4 bits in one round, and a different set of 4 bits in the next round input, and choose some special representation of the field $GF(2^4)$, approximations of the form $\frac{P(X^{2^r})}{Q(X^{2^r})}$ such as as in Theorem 5.2.1 may indeed exist.

**Summary of the General Attack.** We resume here all the different choices that the attacker should explore to find the best attack of this type.

1. Choose the size of the field, for example $m = 4$.
2. Select some input and output $m$-bit masks for one round that can be the same (invariant attacks) or different (much more possibilities). These masks can be subsets of bits, can be linear selection functions and can even by arbitrary non-linear functions $GF(2)^n \to GF(2)^m$.
3. Masks should be selected in such a way that a bias exists from the information theoretical point of view: the output mask seen as a parameterised function of the inputs should not be uniform.
4. Select a representation of the two finite fields.
5. Choose parameters $(D, \varepsilon)$, guess $r$, and find the polynomial equation of Theorem 5.2.1 by a version of Sudan's algorithm, cf. [22].
6. By combining connecting approximations of this type, exactly as in linear cryptanalysis, we obtain attacks for an arbitrary number of rounds.

## 6.1    Our New Attack - Summary of What We Get

1. The possibilities offered by this attack are very large and given a cipher it is hard to know if it can be efficiently applied.

2. It is a special case of GLC. (Looking for attacks that are excessively general doesn't make sense, as we will not be able to explore them and to see if they do apply to a particular cipher).

3. It is not excessively general. Given a particular cipher such as AES, it is hard but probably still possible to find all interesting ways of applying it.

4. It contains linear cryptanalysis. (When $m = 1$. Masks are multivariate functions $GF(2^n) \rightarrow GF(2^1)$, approximation by homographic functions in $GF(2^1)$ amounts to using only the identity function in $GF(2^1)$.).

5. It contains all the attacks of Jakobsen from [22]. It goes beyond: it works locally instead of globally, and does no longer require the components to have low degree approximations.

6. Though both previous attacks will be easily prevented if only we use one big $Inv$ function inside the cipher, the new attack can tolerate an arbitrary number of inverses (of the same size $m$). Thus it allows to cryptanalyse many ciphers that have very high non-linearity and resist to classical attacks.

7. Since it is a generalised linear attack, it is possible to show by using the Fourier Transform, that if the selection of the $m$ bits is done in a linear way, and if $m$ is small, this attack cannot be "much" faster than the best linear attack on the same cipher. It can however be strictly better (and it is obvious to construct examples by embedding noisy polynomials in a round function). In the case of non-linear selection functions, **or** if $m$ is bigger (e.g. $\geq 32$) these attacks can be much faster than any other known attack, as it was already the case in our simple example, cf. point 3a in Section 5.

8. With this attack it is obvious and easy to construct many quite complex ciphers such that their complexity does not grow exponentially with the number of rounds. All we need is to embed in each round an arbitrary combination of the components of Theorem 5.2.1, with the input and the output being hidden. This embedded approximation can be systematically (i.e. for every round) highly non-linear and can also be systematically non-polynomial.

# Part II - Constructing Insecure Feistel Ciphers

## 7   Weak Feistel Ciphers Based on the Inverse S-Box

In this section we will exploit a special case of Bi-Linear Cryptanalysis (BLC) [6], being itself a special case of Generalised Linear Cryptanalysis (GLC) [19].We do not really need to understand the whole BLC [6], and only recall the basic principles of BLC when needed. This paper can be read independently without knowing BLC. BLC allows to construct ciphers that look secure w.r.t. the state of the art in cryptanalysis, yet there are extremely weak. For example, we consider a Feistel cipher in which the round function is given by:

$$f_i(X) = K_i \cdot Inv(X) \quad \text{in } GF(2^n),$$

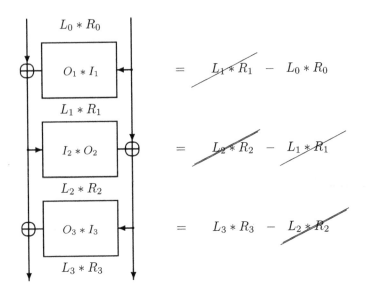

**Fig. 2.** The principle of Bi-linear Cryptanalysis over $GF(2^n)$ for a Feistel cipher

with $K_i \in GF(2^n)$ being the partial key. We will show that this cipher is insecure. Our notations are as follows: We consider a Feistel cipher with $N_r$ rounds. Let $I_i \in GF(2^n)$ and $O_i \in GF(2^n)$ denote respectively the input and the output of the $i$-th round function $i = 1..N_r$, let $(L_0, R_0)$ be the input and $(L_{N_r}, R_{N_r})$ be the output. (Note: in this paper we use "untwisted" version of the Feistel schemes, as on the right-hand figure, page 254 in [26]. Thus the meaning of L and R is as on Fig 2 and differs from several other papers).

We have then (see Fig. 2):

$$L_{N_r} \cdot R_{N_r} \oplus L_0 \cdot R_0 = \sum_{i=1}^{\lceil N_r/2 \rceil} O_{2i-1} \cdot I_{2i-1} \oplus \sum_{i=1}^{\lfloor N_r/2 \rfloor} I_{2i} \cdot O_{2i} = \sum_{i=1}^{N_r} I_i \cdot O_i$$

For every round $i$, by definition of the round function, we have:

$$I_i \cdot O_i = K_i \text{ with probability } \left(1 - \frac{1}{2^n}\right)$$

And thus we get the following I/O sum for the whole cipher:

$$L_{N_r} \cdot R_{N_r} \oplus L_0 \cdot R_0 = \sum_{i=1}^{N_r} K_i \text{ with probability } \left(1 - \frac{1}{2^n}\right)^{N_r}$$

We do not know $\sum K_i$ but it can be recovered from one known plaintext. Then, this equation allows to distinguish our cipher from a random permutation given 2 plaintexts, for $N_r \ll 2^n$ and with negligible error probability of about

$2^{-n}$. Even when the number of rounds is $N_r = 2^n$ our characteristic is true with probability $(1-1/2^n)^{2^n} \approx 1/e$ and thus the cipher can be still distinguished from random with good error probability, given about $\mathcal{O}(e^2)$, i.e. a small constant number of plaintexts. (Our formula allows also to recover the plaintext if a half of it can be guessed in a dictionary attack.)

We get a cipher with the following properties:

1. It is based on the inverse in $GF(2^k)$.
2. It mixes two group operations: addition and multiplication in $GF(2^n)$.
3. It has very good diffusion and avalanche properties.
4. It is composed of very "good" Boolean functions (cf. [4]) and thus resists to all common attacks on block ciphers including LC and DC.
5. Yet the security of it does **not** grow exponentially with the number of rounds. It is easy to break in practice even for $2^n$ rounds, given about 10 plaintexts.

## 7.1    More Weak Feistel Ciphers Based on the Inverse S-Box

What we described here is a very special case of the general family of insecure Feistel ciphers specified in [6]. It is possible to see that in general we have:

**Theorem 7.1.1 (General Construction of Weak Feistel Ciphers).** If the round function is such that there is a symmetric quadratic multivariate relation over $GF(2)$ (or any other field) between the input and output bits, then the cipher is insecure and can be distinguished from a random permutation given a small constant number of plaintexts.

We omit the proof. (This result is new, but rather obvious if we read [6].)

**Further Extensions.** Obviously the equations can be probabilistic. Also the representation of the field may be secret, however here, unlike in Section 6, if this representation is linear over $GF(2)$, it does not help: bi-linear equations over $GF(2^k)$ always give bi-linear equations over $GF(2)$. Then, given a cipher it becomes a hard problem to see if it is weak w.r.t. such attacks, even if we are aware of Theorem 7.1.1. Moreover weak ciphers do not limit to these specified by this theorem. In general the multivariate bi-linear relations can also include some linear parts that, when combined for a whole cipher, can be recovered from linear cryptanalysis. Below we such example, that really does not look as a weak cipher, yet it can be broken easily even for one thousand rounds.

**Example.** We consider a 64-bit Feistel cipher with 32-bit constant parameter $(c, c') \in GF(2^{16})^2$ expanded key being $(K_1, K'_1, \ldots, K_{N_r}, K'_{N_r}$ and in which the round function is $(x, x') \mapsto (y, y')$ with $x, x', y, y' \in GF(2^{16})$ defined as:

$$\begin{cases} y = \frac{x+x'+K_i}{x+c} & \text{in } GF(2^{16}) \\ y' = \frac{x+x'+K'_i}{x'+c'} & \text{in } GF(2^{16}) \end{cases}$$

This cipher looks very secure, yet we have the following symmetric relation with additional linear terms, true with probability about $1 - 2^{-15}$:

$$xy + x'y' = K_i + K'_i + yc + y'c' \quad \text{in } GF(2^{16})$$

This implies that for the whole cipher, if we denote the left part of the input by $(L_0, L'_0)$ and so on respectively, we have in $GF(2^{16})$:

$$L_{N_r} \cdot R_{N_r} \oplus L'_{N_r} \cdot R'_{N_r} \oplus L_0 \cdot R_0 \oplus L'_0 \cdot R'_0 = \sum_{i=1...N_r} (K_i \oplus K'_i) \oplus \sum_{i=1...N_r} (c \cdot O_i \oplus c' \cdot O'_i) =$$

$$= \sum_{i=1...N_r} (K_i \oplus K'_i) \oplus c \cdot (L_0 \oplus L_{N_r} \oplus R_0 \oplus R_{N_r}) \oplus c' \cdot (L'_0 \oplus L'_{N_r} \oplus R'_0 \oplus R'_{N_r}).$$

This equation holds with probability about $(1 - 2^{-15})^{N_r}$ and allows to distinguish the cipher from a random permutation given a few plaintexts, whatever is the number of rounds, for up to about $2^{15}$ rounds. Another very weak cipher based on inverse in $GF(2^n)$.

**Extensions of This Particular Construction.** As in [6] it is possible to see that if $G(x, x')$ is a component (that can be key-dependent) such that fixed some linear combination of outputs of $G$ is biased, then we can replace $K_i$ by $K_i + G(x, x')$ and $K'_i$ by $K'_i + G(x, x')$ in our definition of the round function, and the cipher will still be weak. We can also replace $x + x'$ by an arbitrary function of two variables (the same in both parts).

## 8    Generalised Feistel Ciphers

We can use similar tricks to ciphers similar to SHA or Skipjack, and give many other constructions of insecure ciphers based on the inverse in $GF(2^n)$. We give here an example. We will build a 64-bit block cipher. We divide our 64-bit state in 4 parts $a, b, c, d$ and each round is as follows:

$$\begin{cases} b \leftarrow a \\ c \leftarrow b \\ d \leftarrow c \\ a \leftarrow d + K_i \cdot Inv(a + b + c) \end{cases}$$

Again it looks very good, mixes all kind of operations... and is very weak. It can be broken not exactly by bi-linear cryptanalysis (BLC), but by a more general attack that can be called **Multi-Linear Cryptanalysis (MLC)**. For example, we consider the following expression in $GF(2^{16})$:

$$ab + ac + ad + bc + bd + cd$$

It is symmetric by any permutation of the 4 parts. After one round of encryption the same expression becomes:

$$ab + ac + bc + [d + K_i \cdot Inv(a + b + c)](a + b + c)$$

The difference between the two previous expressions is:

$$K_i \cdot Inv(a + b + c) \cdot (a + b + c)$$

which is equal to $K_i$ with probability close to 1. Again, we can sum up these differences over the whole cipher and this allows to break our cipher given $\mathcal{O}(1)$ plaintexts for a large number of rounds up to about $2^{16}$.

## 8.1    Higher Degree Multi-linear Cryptanalysis

MLC is not limited to quadratic equations. It is easy to show the following result:

**Theorem 8.1.1 (General Principle of Multi-linear Cryptanalysis).**
Consider a cipher with $d$ parts in $GF(2^n)$ in which each round transforms $(a_1, \ldots, a_d) \mapsto (a'_1, \ldots, a'_d) = (a_d \oplus F_K(a_1, \ldots, a_{d-1}), a_1, \ldots, a_{d-1})$ with an arbitrary function $F$. Let $P(a_1, \ldots, a_d)$ be an arbitrary $d$-linear function in $GF(2^n)$. Then for each round we have:

$$P(a_1, \ldots, a_d) - P(a'_2, \ldots, a'_d, a'_1) = P(a_1, a_2, \ldots, a_{d-1}, F_K(a_1, \ldots, a_{d-1}))$$

With this, we can construct multi-linear characteristics for an arbitrary number of rounds of a Feistel cipher. they will be composed of a common (the same for every round) $d$-linear part and some $(d-1)$-linear expressions that connect one round to another.

**Simple Example:** We leave the reader the pleasure to find that the following two ciphers are very easy to break by the new MLC attack:

$$
\begin{cases}
b \leftarrow a \\
c \leftarrow b \\
d \leftarrow c \\
a \leftarrow d + K_i \cdot Inv(abc)
\end{cases}
\qquad
\begin{cases}
b \leftarrow a \\
c \leftarrow b \\
d \leftarrow c \\
a \leftarrow d + K_i \cdot Inv(ab + bc + ac)
\end{cases}
.
$$

**Extensions:** There are many extensions and generalisations possible. We expect that several real-life ciphers such as SHACAL or Skipjack should have interesting attacks of this type. However the number of possible MLC attacks is quite big. and systematic exploration of these attacks will not be obvious achieve.

## 9    Conclusion

Proposing insecure ciphers with highly non-linear components may look as an exercise with no definite purpose. However each time we do so, we can usually formulate a general class of attacks that can potentially be applied to (more or less) any cipher. In this paper we introduced several new types of Generalised Linear Cryptanalysis. The universe of such attacks is unfortunately excessively rich, and remains largely unexplored. We show their interest by constructing various insecure ciphers. Their specific form is not trivial and is determined by the high level structure of the cipher. Locally, they exploit the existence if some non-linear multivariate relations that come from the inverse S-box. This demonstrates that such S-boxes can be dangerous and lead to devastating attacks.

In order to prevent such attacks, we advocate, following [10], to use S-boxes that have no such simple polynomial relations. In particular, for software encryption, we can afford to use reasonably large random S-boxes. This should prevent all known attacks on block ciphers: linear/differential cryptanalysis with generalisations, all kinds of attacks described in this paper, and also any kind of global algebraic attack such as XSL [10].

**Acknowledgments.** We would like to thank Louis Goubin and Lars Knudsen for helpful remarks and encouragements.

# References

1. Frederik Armknecht, Matthias Krause: *Algebraic Atacks on Combiners with Memory*, Crypto 2003, LNCS 2729, pp. 162-176, Springer.
2. Kazuaro Aoki and Serge Vaudenay: *On the Use of GF-Inversion as a Cryptographic Primitive.* SAC 2003, LNCS 3006, pp. 234-247, Springer 2004.
3. Ross Anderson, Eli Biham and Lars Knudsen: *Serpent: A Proposal for the Advanced Encryption Standard.*
4. Anne Canteaut, Marion Videau: *Degree of composition of highly nonlinear functions and applications to higher order differential cryptanalysis*, Eurocrypt 2002, LNCS 2332, Springer.
5. K.W. Campbell, M.J. Wiener: *Proof that DES is not a group.* Crypto'92, LNCS 740, pp. 512-520, Springer-Verlag, New York, 1993.
6. Nicolas Courtois: *Feistel Schemes and Bi-Linear Cryptanalysis*, To be presented at Crypto 2004, Santa Barbara, California, 15-19 August 2004.
7. Nicolas Courtois: *The security of Hidden Field Equations (HFE)*; Cryptographers' Track Rsa Conference 2001, LNCS 2020, Springer, pp. 266-281.
8. Nicolas Courtois: *Algebraic Attacks on Combiners with Memory and Several Outputs*, Available on `http://eprint.iacr.org/2003/125/`. 23 June 2003.
9. Nicolas Courtois: *Fast Algebraic Attacks on Stream Ciphers with Linear Feedback*, Crypto 2003, LNCS 2729, pp: 177-194, Springer.
10. Nicolas Courtois and Josef Pieprzyk, *Cryptanalysis of Block Ciphers with Overdefined Systems of Equations*, Asiacrypt 2002, LNCS 2501, pp.267-287, Springer, a preprint with a different version of the attack is available at `http://eprint.iacr.org/2002/044/`.
11. Nicolas Courtois: *Higher Order Correlation Attacks, XL algorithm and Cryptanalysis of Toyocrypt*, ICISC 2002, LNCS 2587, pp. 182-199, Springer.
12. Nicolas Courtois and Willi Meier: *Algebraic Attacks on Stream Ciphers with Linear Feedback*, Eurocrypt 2003, Warsaw, Poland, LNCS 2656, pp. 345-359, Springer. An extended version is available at `http://www.minrank.org/toyolili.pdf`
13. Nicolas Courtois, Magnus Daum and Patrick Felke: *On the Security of HFE, HFEv- and Quartz*, PKC 2003, LNCS 2567, Springer, pp. 337-350. The extended version can be found at `http://eprint.iacr.org/2002/138/`.
14. Joan Daemen, Vincent Rijmen: *AES proposal: Rijndael*, `http://csrc.nist.gov/encryption/aes/rijndael/Rijndael.pdf`
15. Joan Daemen, Vincent Rijmen: *The Design of Rijndael. AES - The Advanced Encryption Standard*, Springer-Verlag, Berlin 2002. ISBN 3-540-42580-2.
16. Joan Daemen, Vincent Rijmen, Bart Preneel, Anton Bosselaers, Erik De Win: *The Cipher SHARK*, FSE 1996, Springer.

17. Niels Ferguson, Richard Schroeppel and Doug Whiting: *A simple algebraic representation of Rijndael,* SAC 2001, page 103, LNCS 2259, Springer.

18. Lars Knudsen: *Block Ciphers - Analysis, Design and Applications,* PhD thesis, Aarhus University, Denmark, 1994.

19. C. Harpes, G. Kramer, and J. Massey: *A Generalization of Linear Cryptanalysis and the Applicability of Matsui's Piling-up Lemma,* Eurocrypt'95, LNCS 921, Springer, pp. 24-38. `http://www.isi.ee.ethz.ch/ harpes/GLClong.ps`

20. Thomas Jakobsen and Lars Knudsen: *Attacks on Block Ciphers of Low Algebraic Degree,* Journal of Cryptology 14(3): 197-210 (2001).

21. Thomas Jakobsen: *Higher-Order Cryptanalysis of Block Ciphers.* Ph.D. thesis, Dept. of Math., Technical University of Denmark, 1999.

22. Thomas Jakobsen: *Cryptanalysis of Block Ciphers with Probabilistic Non-Linear Relations of Low Degree,* Crypto 98, LNCS 1462, Springer, pp. 212-222, 1998.

23. Thomas Jakobsen, Lars R. Knudsen: *The Interpolation Attack on Block Ciphers,* FSE 97, LNCS 1267, Springer, pp.28-40, 1997.

24. Antoine Joux, Jean-Charles Faugère: *Algebraic Cryptanalysis of Hidden Field Equation (HFE) Cryptosystems Using Gröbner Bases,* Crypto 2003, LNCS 2729, pp. 44-60, Springer, 2003.

25. R. Lidl, H. Niederreiter: *Finite Fields,* Encyclopedia of Mathematics and its applications, Volume 20, Cambridge University Press.

26. Alfred J. Menezes, Paul C. van Oorschot, Scott A. Vanstone: *Handbook of Applied Cryptography*; CRC Press, 1996.

27. S. Murphy, M. Robshaw: *Essential Algebraic Structure within the AES,* Crypto 2002, Springer.

28. Kaisa Nyberg: *Differentially Uniform Mappings for Cryptography,* Eurocrypt'93, LNCS 765, Springer, pp. 55-64.

29. Jacques Patarin: *Hidden Fields Equations (HFE) and Isomorphisms of Polynomials (IP): two new families of Asymm. Algorithms,* Eurocrypt'96, Springer, pp. 33-48.

30. Adi Shamir, Jacques Patarin, Nicolas Courtois, Alexander Klimov, *Efficient Algorithms for solving Overdefined Systems of Multivariate Polynomial Equations,* Eurocrypt'2000, LNCS 1807, Springer, pp. 392-407.

31. Claude Elwood Shannon: *Communication theory of secrecy systems,* Bell System Technical Journal 28 (1949).

# A    Proof of Theorem 4.3.1

In this section we give the proof of Theorem 4.3.1:

**Theorem 4.3.1 (The Group Generated by $Inv$ and XORs).**
The group generated by composing $Inv$ and key additions is exactly the group of all permutations of $GF(2^n)$.

**Proof:** Let $K = GF(2^n)$ and let $a \in K - \{0\}$. The following equality that holds for all $X \in \overline{K}$ and any ring/field of characteristic 2:

$$a^2 X = \cfrac{1}{1/a + \cfrac{1}{a + \frac{1}{1/a+X}}}$$

We propose to replace the inverse in $\overline{K}$ by the Rijndael inverse. We get the following function defined for $X \in K$:

$$A_a(X) = Inv\left(Inv(a) \oplus Inv\left(a \oplus Inv\left(Inv(a) \oplus X\right)\right)\right)$$

Again, since $Inv$ and $1/x$ are almost always equal, this function must be equal to $a^2 X$ with overwhelming probability. By inspection we verify that:

$$\begin{cases} A_a(X) = a^2 X \text{ for } X \notin \{0, 1/a\} \\ A_a(1/a) = 0 \\ A_a(0) = a \end{cases}$$

Example: when $a = 1$, we get a function $B = A_1$ that is equal to identity except that it swaps two points 0 and 1:

$$\begin{cases} A_1(X) = X \text{ for } X \notin \{0, 1\} \\ A_1(1) = 0 \\ A_1(0) = 1 \end{cases}$$

We will construct more functions that exchange points. Let $a \notin \{0, 1\}$. We define:

$$C_a(X) = A_a(A_1(A_{\frac{1}{a}}(X)))$$

By inspection we verify that this function exchanges exactly two points $a$ and $a^2$:

$$\begin{cases} C_a(X) = X \text{ for } X \notin \{a, a^2\} \\ C_a(a) = a^2 \\ C_a(a^2) = a \end{cases}$$

For the next step, we observe that for any field $K = GF(2^n)$ the square is a permutation and $\sqrt{a}$ is well defined. We define the following function:

$$D_a(X) = A_{\frac{1}{\sqrt{a}}}(C_a(A_{\sqrt{a}}(X)))$$

By inspection we verify that for $a \notin \{0, 1\}$ this function exchanges exactly two points 1 and $a$:

$$\begin{cases} D_a(X) = X \text{ for } X \notin \{1, a\} \\ D_a(1) = a \\ D_a(a) = 1 \end{cases}$$

We verified that it remains true also for $a = 0$ (otherwise we could use $B = A_1$ to exchange 0 and 1). Thus we can exchange 1 and any other point. Finally we can exchange any couple of points $(a, b)$ as follows:

$$\begin{cases} E_{ab} = D_a \circ D_b \circ D_a \text{ for } a \neq 1, b \neq 1 \\ E_{1a} = E_{a1} = D_a \text{ for } a \neq 1. \end{cases}$$

The transformations $E_{ab}$ generate the group of all permutations.    □

# Author Index

# Lecture Notes in Computer Science

For information about Vols. 1–3476

please contact your bookseller or Springer

Vol. 3523: J.S. Marques, N. Pérez de la Blanca, P. Pina (Eds.), Pattern Recognition and Image Analysis, Part II. XXVI, 733 pages. 2005.

Vol. 3522: J.S. Marques, N. Pérez de la Blanca, P. Pina (Eds.), Pattern Recognition and Image Analysis, Part I. XXVI, 703 pages. 2005.

Vol. 3521: N. Megiddo, Y. Xu, B. Zhu (Eds.), Algorithmic Applications in Management. XIII, 484 pages. 2005.

Vol. 3520: O. Pastor, J. Falcão e Cunha (Eds.), Advanced Information Systems Engineering. XVI, 584 pages. 2005.

Vol. 3519: H. Li, P. J. Olver, G. Sommer (Eds.), Computer Algebra and Geometric Algebra with Applications. IX, 449 pages. 2005.

Vol. 3518: T.B. Ho, D. Cheung, H. Liu (Eds.), Advances in Knowledge Discovery and Data Mining. XXI, 864 pages. 2005. (Subseries LNAI).

Vol. 3517: H.S. Baird, D.P. Lopresti (Eds.), Human Interactive Proofs. IX, 143 pages. 2005.

Vol. 3516: V.S. Sunderam, G.D.v. Albada, P.M.A. Sloot, J.J. Dongarra (Eds.), Computational Science – ICCS 2005, Part III. LXIII, 1143 pages. 2005.

Vol. 3515: V.S. Sunderam, G.D.v. Albada, P.M.A. Sloot, J.J. Dongarra (Eds.), Computational Science – ICCS 2005, Part II. LXIII, 1101 pages. 2005.

Vol. 3514: V.S. Sunderam, G.D.v. Albada, P.M.A. Sloot, J.J. Dongarra (Eds.), Computational Science – ICCS 2005, Part I. LXIII, 1089 pages. 2005.

Vol. 3513: A. Montoyo, R. Muñoz, E. Métais (Eds.), Natural Language Processing and Information Systems. XII, 408 pages. 2005.

Vol. 3512: J. Cabestany, A. Prieto, F. Sandoval (Eds.), Computational Intelligence and Bioinspired Systems. XXV, 1260 pages. 2005.

Vol. 3510: T. Braun, G. Carle, Y. Koucheryavy, V. Tsaoussidis (Eds.), Wired/Wireless Internet Communications. XIV, 366 pages. 2005.

Vol. 3509: M. Jünger, V. Kaibel (Eds.), Integer Programming and Combinatorial Optimization. XI, 484 pages. 2005.

Vol. 3508: P. Bresciani, P. Giorgini, B. Henderson-Sellers, G. Low, M. Winikoff (Eds.), Agent-Oriented Information Systems II. X, 227 pages. 2005. (Subseries LNAI).

Vol. 3507: F. Crestani, I. Ruthven (Eds.), Information Context: Nature, Impact, and Role. XIII, 253 pages. 2005.

Vol. 3506: C. Park, S. Chee (Eds.), Information Security and Cryptology – ICISC 2004. XIV, 490 pages. 2005.

Vol. 3505: V. Gorodetsky, J. Liu, V. A. Skormin (Eds.), Autonomous Intelligent Systems: Agents and Data Mining. XIII, 303 pages. 2005. (Subseries LNAI).

Vol. 3504: A.F. Frangi, P.I. Radeva, A. Santos, M. Hernandez (Eds.), Functional Imaging and Modeling of the Heart. XV, 489 pages. 2005.

Vol. 3503: S.E. Nikoletseas (Ed.), Experimental and Efficient Algorithms. XV, 624 pages. 2005.

Vol. 3502: F. Khendek, R. Dssouli (Eds.), Testing of Communicating Systems. X, 381 pages. 2005.

Vol. 3501: B. Kégl, G. Lapalme (Eds.), Advances in Artificial Intelligence. XV, 458 pages. 2005. (Subseries LNAI).

Vol. 3500: S. Miyano, J. Mesirov, S. Kasif, S. Istrail, P. Pevzner, M. Waterman (Eds.), Research in Computational Molecular Biology. XVII, 632 pages. 2005. (Subseries LNBI).

Vol. 3499: A. Pelc, M. Raynal (Eds.), Structural Information and Communication Complexity. X, 323 pages. 2005.

Vol. 3498: J. Wang, X. Liao, Z. Yi (Eds.), Advances in Neural Networks – ISNN 2005, Part III. XLIX, 1077 pages. 2005.

Vol. 3497: J. Wang, X. Liao, Z. Yi (Eds.), Advances in Neural Networks – ISNN 2005, Part II. XLIX, 947 pages. 2005.

Vol. 3496: J. Wang, X. Liao, Z. Yi (Eds.), Advances in Neural Networks – ISNN 2005, Part II. L, 1055 pages. 2005.

Vol. 3495: P. Kantor, G. Muresan, F. Roberts, D.D. Zeng, F.-Y. Wang, H. Chen, R.C. Merkle (Eds.), Intelligence and Security Informatics. XVIII, 674 pages. 2005.

Vol. 3494: R. Cramer (Ed.), Advances in Cryptology – EUROCRYPT 2005. XIV, 576 pages. 2005.

Vol. 3493: N. Fuhr, M. Lalmas, S. Malik, Z. Szlávik (Eds.), Advances in XML Information Retrieval. XI, 438 pages. 2005.

Vol. 3492: P. Blache, E. Stabler, J. Busquets, R. Moot (Eds.), Logical Aspects of Computational Linguistics. X, 363 pages. 2005. (Subseries LNAI).

Vol. 3489: G.T. Heineman, I. Crnkovic, H.W. Schmidt, J.A. Stafford, C. Szyperski, K. Wallnau (Eds.), Component-Based Software Engineering. XI, 358 pages. 2005.

Vol. 3488: M.-S. Hacid, N.V. Murray, Z.W. Raś, S. Tsumoto (Eds.), Foundations of Intelligent Systems. XIII, 700 pages. 2005. (Subseries LNAI).

Vol. 3486: T. Helleseth, D. Sarwate, H.-Y. Song, K. Yang (Eds.), Sequences and Their Applications - SETA 2004. XII, 451 pages. 2005.

Vol. 3483: O. Gervasi, M.L. Gavrilova, V. Kumar, A. Laganà, H.P. Lee, Y. Mun, D. Taniar, C.J.K. Tan (Eds.), Computational Science and Its Applications – ICCSA 2005, Part IV. LXV, 1362 pages. 2005.

Vol. 3482: O. Gervasi, M.L. Gavrilova, V. Kumar, A. Laganà, H.P. Lee, Y. Mun, D. Taniar, C.J.K. Tan (Eds.), Computational Science and Its Applications – ICCSA 2005, Part III. LXV, 1340 pages. 2005.

Vol. 3481: O. Gervasi, M.L. Gavrilova, V. Kumar, A. Laganà, H.P. Lee, Y. Mun, D. Taniar, C.J.K. Tan (Eds.), Computational Science and Its Applications – ICCSA 2005, Part II. LXV, 1316 pages. 2005.

Vol. 3480: O. Gervasi, M.L. Gavrilova, V. Kumar, A. Laganà, H.P. Lee, Y. Mun, D. Taniar, C.J.K. Tan (Eds.), Computational Science and Its Applications – ICCSA 2005, Part I. LXV, 1234 pages. 2005.

Vol. 3479: T. Strang, C. Linnhoff-Popien (Eds.), Location- and Context-Awareness. XII, 378 pages. 2005.

Vol. 3478: C. Jermann, A. Neumaier, D. Sam (Eds.), Global Optimization and Constraint Satisfaction. XIII, 193 pages. 2005.

Vol. 3477: P. Herrmann, V. Issarny, S. Shiu (Eds.), Trust Management. XII, 426 pages. 2005.